YALE AGRARIAN STUDIES SERIES

James C. Scott, series editor

The Agrarian Studies Series at Yale University Press seeks to publish out-standing and original interdisciplinary work on agriculture and rural society —for any period, in any location. Works of daring that question existing paradigms and fill abstract categories with the lived experience of rural peo-ple are especially encouraged.

JAMES C. SCOTT, *Series Editor*

James C. Scott, *Seeing Like a State: How Certain Schemes to Improve the Human Condition Have Failed*

Steve Striffler, *Chicken: The Dangerous Transformation of America's Favorite Food*

James C. Scott, *The Art of Not Being Governed: An Anarchist History of Upland Southeast Asia*

Parker Shipton, *Credit Between Cultures: Farmers, Financiers, and Misunderstanding in Africa*

Sara M. Gregg, *Managing the Mountains: Land Use Planning, the New Deal, and the Creation of a Federal Landscape in Appalachia*

Michael R. Dove, *The Banana Tree at the Gate: A History of Marginal Peoples and Global Markets in Borneo*

Edwin C. Hagenstein, Sara M. Gregg, and Brian Donahue, eds., *American Georgics: Writings on Farming, Culture, and the Land*

Timothy Pachirat, *Every Twelve Seconds: Industrialized Slaughter and the Politics of Sight*

Andrew Sluyter, *Black Ranching Frontiers: African Cattle Herders of the Atlantic World, 1500–1900*

Brian Gareau, *From Precaution to Profit: Contemporary Challenges to Environmental Protection in the Montreal Protocol*

Kuntala Lahiri-Dutt and Gopa Samanta, *Dancing with the River: People and Life on the Chars of South Asia*

Alon Tal, *All the Trees of the Forest: Israel's Woodlands from the Bible to the Present*

Felix Wemheuer, *Famine Politics in Maoist China and the Soviet Union*

Jenny Leigh Smith, *Works in Progress: Plans and Realities on Soviet Farms, 1930–1963*

For a complete list of titles in the Yale Agrarian Studies Series, visit www .yalebooks.com/yupbooks/seriespage.asp?series-94.

Works in Progress

Plans and Realities on Soviet Farms, 1930–1963

JENNY LEIGH SMITH

Yale UNIVERSITY PRESS   New Haven and London

Published with assistance from the foundation established in memory
of James Wesley Cooper of the Class of 1865, Yale College.

Yale University Press books may be purchased in quantity for educational, business,
or promotional use. For information, please e-mail sales.press@yale.edu
(U.S. office) or sales@yaleup.co.uk (U.K. office).

Set in Minion type by Integrated Publishing Solutions.
Printed in the United States of America.

ISBN: 978-0-300-20069-0

Library of Congress Control Number: 2014942184

A catalogue record for this book is available from the British Library.

This paper meets the requirements of ANSI/NISO Z39.48-1992
(Permanence of Paper).

10 9 8 7 6 5 4 3 2 1

Contents

Acknowledgments vii

Introduction 1
1 Model Farms and Foreign Experts 21
2 Restoring Control 63
3 Animal Farms 106
4 Substituting Meat 151
5 The Old and the New 188
Epilogue 225

List of Archives and Contemporary Periodicals 233
Notes 235
Glossary of Russian Terms 256
Index 259

Acknowledgments

There are a great many people and institutions that deserve my sincerest thanks for help with this project; without them, this book would have been impossible. I can only thank a fraction of them here. When this project was in its earliest stages, I received excellent feedback from colleagues at MIT including Deborah Fitzgerald, Shane Hamilton, Dave Luscko, Sara Pritchard, Harriet Ritvo, William Turkel, Elizabeth Wood, Christopher York, and Anya Zilberstein. Having Deborah Fitzgerald as a mentor in particular was pure pleasure. In Moscow, Larisa Petrovna Belozerova, Olga Elina, Lada Lekai, Alexander Nikulin, and Teodor Shanin were all incredibly generous with their time, and patient with my Russian. In St. Petersburg, Alexandra Bekasova and Julia Lajus have been wonderful colleagues. At the Royal Institute of Technology in Stockholm, where I wrote much of the first draft of this work, Maja Fjaestad, Arne Kaijser, Per Lundin, and Brita Lundstrom made me feel right at home. At Yale University, Lloyd Ackert, Valentine Cadieux, Honor Sachs, Jim Scott, and John Varty all offered encouragement, advice, and occasional libation. Dan Kevles in particular was a fantastic mentor at Yale. At my current institution, Georgia Tech, Jonathan Schneer read and commented on the

entire manuscript, and John Krige read it twice. I am indebted
to both of them. Outside of formal institutional affiliations, William DeJong-Lambert, Elizabeth Dunn, Loren Graham, Bar-
bara Hahn, Martha Lampland, Diana Mincyte, and Katherine
Verdery have all helped shape this project, and I am grateful
for their input. What might have been a long, bleak year in
Moscow was made delightful by a number of fellow travelers,
especially Alan Barenberg, Joy Johnston, Timothy Johnston,
Sonja Schmid, and Ben Tromly. Jake Rudnitsky's decades-long
open door policy has saved me thousands of dollars and de-
serves its own category of gratitude.

My initial research in Russia was funded by a grant from
the International Research and Exchanges Board. Additional research time in Iowa, Russia, and Ukraine was supported by NSF
Dissertation Improvement Grant #SES-0449767. The George
Kennan Institute of the Woodrow Wilson Center supported a
summer of research in Washington, and I am grateful to the
many resources the institute made available to me. Later trips
to Russia, as well as time to write and revise this book, were
generously provided first by Yale University and then by Georgia Institute of Technology, and I am grateful for their support.

This project also had significant help from a variety of
librarians and other research professionals. In Moscow, at the
Russian State Archive of the Economy I am most grateful to
the heads of the reading room. In the United States, I worked
at the Library of Congress, especially the Slavic Division of the
European Reading Room. At Harvard University Susan Gar-
dos was incredibly helpful. At Iowa State University, Tanya
Zanish-Belcher was an excellent resource. At the University of
Illinois at Urbana Champaign, Jan Adamczyk was very helpful. Finally, Guy Bush Jr., whose father figures prominently in
the first chapter of this book, contacted me two years ago and

generously transcribed and e-mailed me every letter his father had sent to his mother while he lived and worked in the Soviet Union. These personal letters were invaluable to my research, and I am grateful for his time and care with this project. At Yale University Press, it has been a privilege to work with Jean Thomson Black, and I am grateful for thoughtful feedback from two anonymous reviewers who helped make this book better.

Finally, my family, and in particular my husband, Albert Wan, showed incredible patience and fortitude during this whole project for which I am very, very grateful. Thanks, babe.

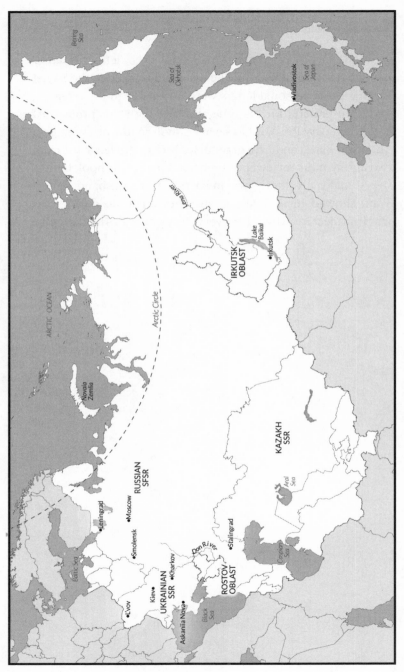

Map 1. The Soviet Union after World War II

Introduction

This book is about the distance between plans and reality in the Soviet countryside during a period (1930–63) when the Soviet Union sought to industrialize and modernize its agricultural system. Most histories of Soviet farms emphasize how badly this project failed—correctly pointing out that for the entire Soviet period, agriculture was inefficient, unpredictable, and environmentally devastating. My own inquiry emerged from the observation that these failures were hardly unique to the Soviet Union but are common to all industrial agricultural systems. Of course, Soviet planners did not set out to create a wasteful, polluting, and unreliable system of agriculture. The Soviet government believed that by modernizing its agricultural system the country would dramatically increase agricultural productivity, relieve millions of rural workers from the tedium of manual labor, and create a model, sustainable system of socialist agriculture that could be imitated in other countries around the world. This is not what happened, but the gap between plans and reality is an interesting story, worthy of a more restrained and thoughtful analysis than historians have previously devoted to the topic.

Did the Soviet Union fall especially short in achieving its goals for a modern industrialized agricultural system? Was the gap between plan and reality in the Soviet Union much greater than this same disconnect in a democratic, capitalist country like the United States? I argue here that there is not a simple yes or no answer to this question. Sometimes, the Soviet Union's agricultural reforms failed spectacularly and in ways that were uniquely Soviet; sometimes they succeeded in the face of overwhelming odds. Often, what looked like failure from the outside was actually an effective but unusual stopgap method of solving a problem that was peculiar to the Soviet system.

Historical Background

This history of Soviet farms begins in the year after Stalin's devastating 1930 collectivization drive and ends in 1963, just after the Soviet Union's first major purchase of wheat on the world market, which effectively ended the country's experiments in agricultural self-sufficiency. In the nearly thirty-five years in between, the Soviet Union went from a mostly undeveloped backwater with a weak and dispersed government to an industrial powerhouse, battling the United States for the conquest of space, governed by a central state authority that was always controlling and sometimes cruel. Power and socialist utopianism combined during this period to create a new kind of state, a new relationship between farms and cities, and a new, putatively socialist form of rural existence.

The beginnings of this transformation were humble, and did not indicate their future importance. The Bolsheviks who seized power in Russia in 1917 inherited a diffuse territory that stretched from the Pacific Ocean to the Gulf of Finland, encompassing ten time zones and almost twenty million square kilometers of land. It was the world's largest country by territory, and the most sparsely settled. The landscape was a patchwork of forests, tundra, and marginal farms. While the suburban regions of Moscow, Petrograd, and a few other cities had nascent factory industries, over 90 percent of Soviet citizens were farmers. Draft animals, although scarce in Russia by European standards, outnumbered adult males by a ratio of at least two to one. On the eve of the First World War, only a little over half the Russian population was literate. Russian villages, with their shaggy oxen, barefoot children, and muddy roads, were the defining features of the new Soviet landscape, albeit one unfamiliar to most Bolshevik revolutionaries.[1]

Communism was an urban ideology and its Bolshevik followers were mainly well-read city dwellers. Vladimir Lenin, writing in 1905, expressed a condescending sympathy for the countryside, calling the peasants of the Moscow region "ignorant and conservative . . . paupers," who were doomed to lag at least one, possibly two revolutions behind the more advanced political uprisings he envisioned for the Soviet Union's urban populace.[2] Soviet farmers may have lagged behind their class-conscious urban counterparts, but the Bolsheviks could not afford to ignore their central role as food suppliers for the major cities of Moscow and Petrograd during the Civil War. Famines struck at the heartlands of Russia and Ukraine in 1921 and 1922. Bread was rationed in the cities and support for the new regime wavered as hunger replaced optimism. After the Civil War ended, the earliest agricultural policies of the Soviet Union encouraged private production. The eight years that followed the Civil War were a time of respite and rebuilding for both cities and farms. While urban areas experimented with avant-garde art and architectural styles during the 1920s, the Soviet countryside struggled to recover the agricultural production levels of the era before the First World War.

These relatively lax rural policies ended in 1929 with a mass collectivization drive. For Soviet planners, collectivization was the first step toward agricultural modernization. Modernization also meant adopting new machines, scaling up agriculture by creating larger fields, applying new, efficient techniques to work in fields and barns, and devising new patterns for collecting and distributing food and other farm products. Government planners grossly underestimated how unpopular collectivization would be in the villages. As outsiders, they imagined that collective agriculture was a modern and efficient solution that would gain rapid social acceptance. Instead, farmers saw collectivization as a new form of bond labor where the state once again assumed ownership over all forms of property and imposed labor discipline through force and terror. For much of the 1930s, the Soviet state engaged in a war against the countryside in which the state employed brutal force to maintain and increase its control over the agrarian sector.

Between 1941 and 1945, the Second World War devastated the country. By any estimate the Soviet Union endured more loss than any other nation that fought in this war. Three major cities—Stalingrad, Leningrad, and Kursk—were nearly razed by fighting. One in five adult men did not return from the war, and another 10 percent of the population returned with permanent physical disabilities. As a defense measure, the Red Army burned or otherwise destroyed crops, machines, and buildings as they retreated across western Russia and Ukraine, ruining some of the most productive lands in the country, and much of the agricultural system needed complete rebuilding after the war. This allowed state planners to erase or alter some of their early mistakes in the countryside, but postwar reconstruction was easier to plan than it was to carry out. Just as in the prewar era, state ambitions greatly exceeded the capabilities of the countryside upon which these plans were imposed.

Stalin died in 1953, and Nikita Khrushchev succeeded him as premier in 1956. Khrushchev had great confidence that his personal knowledge of village life and long experience working in the Ministry of Agriculture would help him revolutionize Soviet farming. He bragged that he expected the USSR to "catch up and overtake" the United States' behemoth farm system in a single generation. Khrushchev's plans for reform and improvement focused particularly on increasing the number of farm animals in the Soviet Union. He was convinced that if the country could make more meat and milk available, diets would improve and people would become happier and healthier.

Ultimately Khrushchev's years of enthusiastic reform yielded mixed results. On one hand, he had an excellent advantage coming to power when he did, since agricultural productivity was low and had been slow to recover after the war. Khrushchev was able to pick up the pace and quality of agriculture and this created enthusiasm and a spirit of experimentation in local and national government agencies. On the other hand, Khrushchev's most famous reforms were spectacular failures. His most infamous agricultural project, the Virgin Lands Campaign, involved plowing up uncultivated grasslands in Kazakhstan and Siberia and planting a variety of crops ill-suited to the harsh, dry climate of these regions. Within a few years, the fertile topsoil beneath these grasslands had eroded, stalling the campaign. Any soil scientist might have predicted this outcome, but Khrushchev foolishly trusted that Soviet science would solve the problem of soil erosion before the rich topsoil of the steppes started to blow away. He was wrong.

This was not the only public failure of Khrushchev's tenure, and in 1964 he was forced out as premier. The year before he left, he approved a deal to purchase wheat from the United States, the first time the Soviet Union had bought grain abroad since before the Second World War. Although the Soviet Union did not depend on foreign grain purchases regularly until 1972, this was symbolically the end of an era in which the Soviet Union strived for agricultural self-sufficiency. The post-Khrushchev era shifted away from the grandiose dreams of the early Soviet period, and toward more sober, realistic goals for the farm system.

Soviet Progress

In treating the eventful years from the early days of collectivization under Stalin to the twilight of Khrushchev's regime, I focus both on what the state hoped to accomplish in its agrarian reforms and why it fell short of those goals. In the majority of cases, Soviet agricultural reforms and interventions were not complete failures; instead they were case studies of simultaneous triumph and disappointment. This mixed legacy of Soviet agriculture has not been a common trope in the history of the Soviet Union because Soviet dissidents and refugees, Cold War–era historians, and social scientists whose anti-Soviet biases are evident have written many of the best works about agriculture and the rural Soviet Union.[3] These biased accounts need updating; the Cold War has ended and the Soviet Union no longer exists. The legacy of socialist agricultural modernization remains relevant today, since all modern countries struggle to balance the inefficiencies, environmental degradations, and labor injustices of industrialized agriculture against the relative luxury of cheap, plentiful food. What can the Soviet experience of agricultural industrialization, which featured some of the most ambitious and creative ideas for expanding and streamlining the food supply of the world's largest nation, teach the rest of the world about how (and how not) to modernize farms? That is the question at the heart of my inquiry.

Progress was not a fixed concept for the Soviets. In 1930, visions of Soviet progress and modernity were overrun with airplanes, tractors, electric lighting, and massive factories. For the first generation of Soviet reformers, the revolution would be mechanized. Soviet planners were focused on overcoming cultural and technological backwardness, especially in rural areas. Traditional forms of agriculture, labor organization, and handicraft were discarded in favor of practices the new Bolshevik authority judged to be scientific, modern, and efficient. This desire to improve the countryside endured throughout the Soviet period, but its methods of achieving reform and purging the countryside of its perceived backwardness changed over time.[4]

To accomplish these goals during the 1920s and 1930s, the Bolsheviks looked to industrial efficiency as it was practiced in the United States. Henry Ford and Frederick Taylor may have been notorious capitalists in their home countries, but to Soviet planners, their innovations in management and mass production transcended economic ideologies. In the early twentieth century both socialism and capitalism focused on maximizing efficiency and scaling up production, and Soviet textbooks and articles that profiled Ford and Taylor ignored their business philosophies and concentrated instead on their achievements as managers and efficiency experts. The admiration was mutual. Henry Ford in particular took great interest in Stalin's industrialization drive and the Soviet Union purchased over twenty-five thousand of his company's Fordson tractors between 1921 and 1927. Ford also helped found the Soviet Union's first tractor factory outside of Leningrad in 1924. His business relationship with the Soviets soured in the 1930s, but not before his tractor designs inspired the first generation of Soviet-built farm machines.

After the Second World War, notions of Soviet progress and modernization shifted to focus on human control over nature. Machines remained a part of the equation that would help the Soviet Union master its natural resources, but they were no longer the primary symbol of progress. The Soviet Union had sponsored industrial development projects during the 1930s that included elements of harnessing the power of nature, but during the postwar period controlling nature became the central image of Soviet modernity. Hydroelectric dams, projects to cultivate the so-called Virgin Lands of Kazakhstan and southern Siberia, and new deep-shaft ore mines were all high-profile postwar projects that linked modernization and development to mastering nature. Perhaps most representative of the way in which control over nature and modernity melded in the postwar era was the Stalin Plan for the Transformation of Nature. Beginning in 1948, this project created windbreaks and shelter belts across the Soviet Union by planting trees both to restore old forests and to create new wooded areas. While the Soviet Union successfully planted millions of trees, it struggled to keep these trees alive, and the program was eventually canceled.[5] The Soviet postwar fascination with controlling nature also led Soviet planners to support the flawed theories of Trofim Lysenko, largely because his simplistic plans for improving crops, animals, and forests assumed that the Soviet state was already in control of nature, and that socialism, as the most scientific and humanistic form of statecraft, would make the most efficient use of nature possible, thereby maximizing nature's productivity.

After the Second World War, the Cold War also influenced how the Soviet Union defined progress. Soviet development and rebuilding focused on besting the United States whenever possible. More than simply a race to the top, however, the Cold War also influenced how Soviet planners thought about developing their nation's resources. Faced with limited funds and manpower, Soviet planners competed strategically with the United States in a few key sectors. The space race and the nuclear arms race between the two countries are probably the two most famous competition venues, but agriculture was also an important front of Cold War rivalry for the Soviet Union. The Cold War played a major role in ensuring that the Soviet Union developed its own processes of modernizing and did not simply mimic the United States, as it had done during the 1930s. The farms of the postwar Soviet Union were a crucial Soviet home front of the Cold War that shows how international politics shaped domestic policy and how these events affected the everyday lives of people far removed from positions of power and authority. The farms that the Soviet Union built and rebuilt in the postwar era were more successful than their 1930s predecessors had been and they succeeded in raising the living and dietary standards of Soviet citizens dramatically between 1945 and 1963. Furthermore, rural modernization did not result in a monotonous or soulless rural landscape, minutely and unimaginatively planned from above. Soviet farms had a significant run of success in this period, but they were simultaneously inefficient sites of industrial activity in which waste and pollution continually threatened to overwhelm the benefits that these institutions also yielded. Ultimately, the industrial farming system that the Soviet state created and improved upon during the Cold War proved unsustainable in the long run. In this respect, Soviet agriculture resembled its non-Soviet counterparts in other parts of the world.

During the twentieth century, works toward agricultural progress had a few important goals, regardless of where these projects took place. Industrial systems of agriculture strived to produce more food, to produce better food, and to allow people to eat higher up on the food chain, consuming more animals and animal products. In its quest for a modern food system, the Soviet Union embraced these goals, although the ways in which it realized them were often unique. The Soviet Union struggled to feed itself; the state leaned heavily on the cash that wheat exports brought into the country, and this hampered agricultural autonomy. In every famine year during the Soviet period, the state also exported significant amounts of wheat. For the Soviet Union, a plentiful supply of grain was important for its value on the world market as well as its ability to sustain life. Sometimes wheat's role as a commodity superseded its value as a domestic staple, and food shortages resulted. The state was also interested in producing more varied crops, experimenting with new, higher-yielding plant varieties, and industrializing the production of foods that had previously only been homegrown or gathered wild. To this end, collective farms worked closely with experiment stations to try out new varieties of grains and to introduce more productive, purebred animals into Soviet herds. They also experimented with raising animals like nutrias, foxes, and seals on farms, and domesticating berries and mushrooms.

The most radical modernizing act of Soviet agricultural planners during this time was their move to increase meat and milk production. Nutritionists during the twentieth century often recommended a high-protein diet rich in animal products to fuel industrial laborers. The Soviet state embraced this suggestion and turned it into official state policy. Officials believed that by increasing the number of animal products Soviet citizens ate, the health of the nation would improve. There was little notion during this period that a diet high in animal products was less than optimal. Turning animal agriculture into state policy was by no means a singularly Soviet pursuit; in fact, the Soviet Union directly emulated the United States in its goals to produce more meat and milk; targets for meat and milk production were set specifically in order to surpass the United States in this sector. The Soviet Union's cold environment, vast geography, and weak legacy of animal agriculture all posed challenges to the goal of increasing meat and milk production. While the Soviet Union's achievements in animal agriculture fell short of its goals, its achievements were not trivial.

Agriculture and the Environment

This book pays close attention to the environmental challenges the Soviet Union faced as it modernized and industrialized its agricultural system. Farming has been, after all, the most widespread and dramatic human transformation of the earth's environment. In dealing with the environmental legacy of agriculture in the Soviet Union, I adopt an agroecological perspective, examining how modes of food production, especially in the modern era, have altered and in turn been altered by the many environments of the Soviet Union.[6] The usual observation made about farming in the Soviet Union is that the cold climate caused problems. This was true, but Soviet agricultural planners confronted far greater challenges. Sprawled as it was across the entire northern Eurasian continent, the Soviet Union featured a vast diversity and range of environmental extremes. Soviet plans for a modern system of agriculture originally strived for common policies and universal approaches, but this was not always practical in such an environmentally diverse country. Collectivized farms worked well in some parts of the country, but never made sense in others. Some of the Soviet Union's soils held up well under mechanized plowing, while others eroded after a few years of machine-based cultivation. Conversely, sometimes state visions of progress were too specific rather than too universal. For example, in the 1950s, the Ministry of Agriculture strived to develop plants and animals that were uniquely suited to particular Soviet regional environments. Planning for, managing, and tracking the progress at this intimate level strained the Ministry of Agriculture's bureaucratic capabilities just as much as managing impractically universal megaprojects.

Agricultural modernization significantly altered the Soviet Union's environment, often for the worse. Machines and industrial systems of labor organization intensified production, often placing pressure on marginal environments with fragile ecosystems. Soviet planners rarely anticipated long-term environmental consequences of agricultural modernization, but industrial agriculture has left a permanent legacy across the Soviet Union that includes pollution, erosion, desertification, and soil and water contamination. Soviet agricultural progress was often harmful, but quirks of the Soviet state also prevented or reduced some negative environmental impacts of industrial agriculture. In some cases socialist forms of agriculture served as a buffer against pollution and environmental hazards. For example, the Soviet Union lagged behind the United States and most other developed countries in manufacturing and distributing petroleum-based fertilizers and pesticides. Had they been given the opportunity, Soviet planners almost certainly would have chosen to use more chemicals on their fields, but because these were not available, Soviet soils, waters, and agricultural workers remained far less exposed to toxic chemicals than their capitalist-world counterparts. This same technological gap prevented Soviet livestock from receiving significant amounts of antibiotics and growth hormones during the 1950s and 1960s. Because pharmaceuticals were not commonly available to ward off the many diseases that accompanied the intensification of animal agriculture, Soviet livestock operations were forced to scale back grandiose plans that would have housed tens of thousands of chickens under the same roof, or halved the time from birth to slaughterhouse for pigs. The advantages of such technological lag were not appreciated at the time, but in retrospect some of the ways in which Soviet agriculture remained backward and inefficient have become virtues of ecologically sound systems of agriculture.

A caveat is in order, however; the Soviet Union was the largest, most environmentally and geographically diverse country in the world for most of the twentieth century. Although "the Soviet Union" is often referred to here as a unified whole, and although the agricultural landscapes featured in the book are diverse, the analysis skews toward the historic agricultural heartland of the Soviet Union in European Russia and Ukraine. Some of the topics not discussed extensively, such as collective agriculture in Central Asia and the Virgin Lands project, were omitted because other scholars have written excellent studies of these historic episodes.[7] However, the most common reason for leaving out important and fascinating topics in the history of Soviet agriculture during the middle decades of the twentieth century was simply because there were far too many from which to choose. It was not possible to address every angle of agricultural modernization in the Soviet Union. The case studies that appear here were particularly interesting, and were supported by good archival documentation that addressed an important issue in the legacy of Soviet agricultural modernization, but this analysis is not exhaustive; much has been left out. Many of the arguments in this book about the role of the state in inspiring a unique form of agricultural industrialization would hold true in many other contexts across the Soviet Union, but more examples might help make the story of Soviet agriculture more nuanced and complete. Perhaps other scholars will take up this task.

Overview

In the following chapters, I focus on different times, places, and sets of agricultural modernization goals to illustrate the larger point that state ambitions for modernization and rural improvement often fell short of their main objectives, but that this did not mean the Soviet countryside was not rapidly modernizing, evolving, and adapting to a socialist mentality and new, heavy-handed methods of surveillance and control.

To begin with, the letters and personal reports of two American specialists, Guy Bush and George Heikens, who lived and worked in the Soviet Union in 1930 and 1931, show that the two years immediately after collectivization were a time of frenetic state-sponsored activity, intended to jump-start collective agriculture by using persuasion rather than force. These first years of collectivization were especially challenging. Rural leadership was bad or nonexistent, and newly collectivized farmers, reluctant to give up their autonomy, were poor workers. Lacking homegrown experts, the state hired foreign professionals to help get the new collective farms or kolkhozes off to a good start. Bush and Heikens were two such men. Their letters home echo official and unofficial reports from the post-collectivization countryside that the new kolkhozes were chaotic places to work, and that newly collectivized workers were both confused about their new responsibilities and resentful at having to obey directives from above. The gaps between what the state planned to achieve on its newest, best-funded model collective farms and what it was actually able to accomplish in the first two years after collectivization are a testament to how little control or understanding of rural areas the Soviet government had in these early days of state-run agriculture.

In the second chapter, I turn to a humble food source, the potato; its various viral, bacterial, and insect predators; and the very real threat of famine in the years after the second World War. During the war potatoes became a staple food for many Soviet farmers, especially in Ukraine. Afterward, potatoes remained an advantageous crop to raise and store, both because they did not require as much labor as traditional grain crops, and because they were easier than grain for farmers to hide from government inspectors. After the war, plant diseases and insects posed real threats to the meager food supply of Eastern Europe. However, potato diseases were hard to spot and even harder to stop; it was notoriously tricky to find or treat potatoes for beetles during this era. The Soviet Ministry of Agriculture seized on this weak link in the food supply as a way to recapture control of agricultural production in the parts of the country that had been out of Soviet control during the war. The state created quarantine stations, initially as a way to control the spread of biological invaders. Surveys and inventories tracked plant diseases, potato beetles, and other crop threats in order to guard against widespread infestation.

In fact, with the potato crops of the 1940s, the Soviet Union dodged a biological bullet; none of the diseases and pests that had blighted American and Western European crops before and during the war appeared commonly in Soviet fields. By 1949, quarantine stations and the Ministry of Agriculture knew there was no immediate threat to potato crops, but the surveillance systems the quarantine station had developed were a good way of keeping tabs on potatoes, a crop farmers used to subsist and to resist Soviet authorities, so these systems remained in place. This resulted in fewer farmers growing potatoes and more farmers going back to growing grain, which was a much more useful commodity for the state to collect and sell.

Surveillance was not the only science employed by the state in order to best husband its agricultural resources. Between 1948 and 1964, Soviet agricultural research was influenced by the theories of Trofim Lysenko, a modern-day Lamarckian who eschewed the emerging science of genetics and heredity. By the standards of any era, Lysenko's scientific theories were insupportable, but Lysenko was successful at encouraging flawed agricultural research and misreading ecologically based evidence for over twenty years. In the third chapter the issue is raised of what results bad science can accomplish for the state by examining how Lysenko's theories worked in the realm of animal breeding during the scientist's heyday in the 1950s.

Surprisingly, Lysenkoist theory did well in the area of animal husbandry. While Lysenkoism failed as a persuasive experimental science, it worked wonders for animal agriculture as a management science. In the 1950s Soviet barns and feedlots did not necessarily need the breakthroughs of in vitro fertilization, cellulose silage, ammonia-based feed additives, or other technologies that were emerging in the United States during this same era. The problems Soviet animal farms faced in 1950 remained the same as those discussed earlier: endemic disease, a lack of antibiotics, not enough veterinarians, an inadequate feed supply, poor hygiene, and disorganized care and feeding schedules. Such problems had plagued European and American agriculture at the turn of the century, and experts in these countries had turned to the efficiency and organizational studies of Frederick Taylor and Henry Ford to reorder animal care. In the Soviet Union, Lysenko's folksy, oversimplified interpretation of animal husbandry was easy to follow and yielded obvious results. The system's unstructured breeding methodology and loose rules on registering and copying foreign breeds increased the diversity of livestock on animal farms. While the productivity of animals remained low because of their small size and limited fecundity, their actual numbers steadily increased every year that Lysenko's recommendations for their care and feeding were in place. Whereas most industrialization projects are thought of as simplifying the landscape and reducing genetic diversity on farms, the "bad science" of Lysenkoism created a breeding program that filled small ecological niches. This runs counter to the fundamental claim about the trend of industrial agriculture toward monolithic and impersonal production systems.

The focus of the next chapter is the Soviet Union's struggles to produce more meat and milk and to make these high-quality (often perishable) foods available to all citizens in spite of a weak and often centrally managed distribution network. Soviet farms experienced significant difficulties in producing more meat and milk products, but retailers, nutritionists, and animal breeders all successfully responded to the shortcomings of animal production and helped bridge the gap between the impractical goals of the Soviet state and the achievable reality. These experts recognized the chronic shortages of meat and milk in the market for what they were, productive capacity limits that were not likely to be solved with a single Five Year Plan. These professional intermediaries invented ways to improve nutrition and market a range of processed food products that reflected the scarcity of supply as well as creativity in content. Soviet farm policy may have been striving toward an unreachable goal, but Soviet food policies took more realistic approaches to product development and distribution networks. At a time when American consumers were learning to buy more food (milk in gallon jugs and standard cuts of meat wrapped in cellophane, for example), Soviet consumers were being presented with a greater range of new "substitution" products like soy protein, milk-based sausages, sunflower oil, and tinned fish. These foods reflected the strengths of the Soviet Union's environment (wild fish stocks, cropland ideal for sunflowers, and so on) as well as an ongoing scarcity of meat and overall low animal productivity.

In the final chapter, I examine the limits of the Soviet Union's new agriculture and food networks on its Siberian frontier by looking at the co-evolution of agricultural reform and an emergent Cold War fur trade in Irkutsk Oblast in central Siberia. While Soviet reformers initially intended to eliminate hunting as an economic livelihood in Siberia, an ascendant world fur market and a series of spectacular agro-industrial project failures across the region forced the Soviet Union to reexamine the importance of the fur industry, especially in terms of its ability to draw scarce American dollars into the state's coffers. In the late 1950s, impoverished collective farms across the oblast offered their members the chance to increase their wages through hunting, and this program was much more successful than the large and famous development projects such as the Virgin Lands Campaign that the state had introduced to the area earlier in the decade. Hunting was not simply an economic pursuit; it also became a link between the old world of the Russian empire and the new Soviet enterprise, in which the resonant social category of the European hunter was exploited to great advantage by the state.

The book concludes with the Soviet Union's first foreign grain deal, made just before Nikita Khrushchev was forced out of office. The move was widely unpopular in the Soviet Union as it signaled a tacit acknowledgement that the country had been unable to raise enough grain to meet its needs for both humans and for the many new farm animals that were also consuming grain. Ultimately, however, by 1965 the Soviet Union had admitted defeat in the pursuit of agricultural self-sufficiency, but limited success with its overall project of modernizing and industrializing its food system. When the Bolsheviks came to power they inherited a vast, diffuse agricultural network that allowed almost 80 percent of the population to subsist on farms, but provided only a meager surplus that could be sold for cash or used to feed an urban, industrial population. By 1965, these numbers had been nearly reversed and over 70 percent of Soviet citizens lived in cities and towns and worked at nonfarm jobs. The Soviet Union's problematic and chronically underperforming agricultural system has been held up by the rest of the developed world, especially the United States, as a cautionary tale about the dangers of socialism, totalitarianism, and overgovernance. In fact, however, the Soviet approach to modernizing agriculture was, by most measures, surprisingly successful. Things did not always turn out the way Soviet planners intended, but there is a significant difference between creating a flawed and wasteful industrial agricultural system (something every developed country has done to a greater or lesser extent) and creating a failed agricultural system. In 1965 and for the duration of the Soviet period, the Soviet state operated with a flawed but fundamentally functional agricultural system that was inefficient, vulnerable, chaotic, and frustratingly reliant on the natural environment. In other words, Soviet agriculture had much in common with its counterparts in capitalist countries around the world.

1

Model Farms and Foreign Experts

In the history of Soviet agriculture two events stand out: first, a rapid collectivization drive in 1929 and 1930 that forced Soviet farmers into state-run farming cooperatives, and second, a famine in 1932 and 1933 that killed as many as six million people, most of them Ukrainian. Historians usually present collectivization and the subsequent famine as consecutive, related events, but the relatively stable two-year interval between them revealed an important and enduring weakness of the Soviet system; central planning experts could not bridge the gap between their ambitious plans for modernization and the reality of enduring underdevelopment in rural areas. Money, additional labor, publicity, and other external inputs all failed to close this divide. The distance between plan and reality in 1930 was important because it set a tone for the decades to come. Well into the 1950s, the Soviet state struggled to overcome what its officials perceived as a backward rural mentality. Official efforts to overcome this backwardness were continually frustrated. Initially, state officials blamed their problems on Soviet peasants who refused to support modernizing reforms; however, a careful analysis of the situation immediately after collectivization reveals that the natural environment, rather than humans, posed the most significant obstruction to progress. It was the state's inability to understand, anticipate, and respond to the complicated natural environments of Soviet agriculture that thwarted rural progress.[1]

The state's initial efforts at collectivization established a new bureaucracy on farms and in villages and leaned heavily on outside experts to steer the new kolkhozes toward production goals that had been set in Moscow. Outside expertise was not always a welcome addition to the rural milieu of the Soviet Union in the 1930s. Nevertheless, this tactic of relying on outsiders continued a tradition of the Soviet Union looking abroad for solutions to its agricultural problems.[2] As with many top-down improvement schemes, the agricultural reforms that accompanied collectivization only partially succeeded and, in the fall of 1932, frustrated by the slow pace of progress in the countryside and alarmed by the persistence of rural resistance to communism, Stalin and other leaders magnified significant harvest shortages into a famine as an exercise in state terrorism. The famine sent a blunt message to rural areas that the government intended to control agriculture in the Soviet Union regardless of the human or economic cost.

These two events—collectivization and starvation—stand out for good reasons: both were dramatic and had immediate, permanent repercussions. However, after the mass collectivization of 1930 but before the terror of the 1932 famine, collectivized farmers experienced almost two years of relative lack of violence during which the state tried to reform the countryside through peaceful, bureaucratic means. These years of respite were mainly the result of the vocal and wide-ranging protests against collectivization that took place in March and April 1930. Nowhere were the changes in the countryside more pronounced than on the state's new model kolkhozes that had been designed to serve as vanguards of modern agriculture. These model farms received extra money, supplies, and expertise from the central government in order to achieve the goals of modernization in as short a time as possible.

Local and national Communist Party records and the personal letters of two foreign American specialists, Guy Bush and George Heikens, illuminate the distance between the modernizing ambitions of the state and the reality of life in the rural Soviet Union immediately after collectivization. Although providing extra capital and expertise to model farms was intended to boost agricultural productivity, this rarely happened. Instead, both the official reports and the private correspondence of the American experts reveal that these years were a time of frenetic, but largely ineffective, activity on model farms. Outside prescriptions failed because they were not up to the task of modernizing the countryside in one mass effort. The Soviet Union's environments posed enormous challenges to modernization that were difficult for outside bureaucrats to plan for, or effectively address. Organizing winter animal feed, keeping animals out of the mud, and maintaining roads and rail lines were common challenges across the Soviet landscape. Farms in the far north and those in arid regions faced even greater environmental challenges. After the Second World War Joseph Stalin campaigned to transform and control nature, announcing that the Soviet Union needed to rise above environmental limits and harness natural resources. However, in 1930 the

Soviet state had little control over nature on its new collective farms. This lack of control hurt both the state's legitimacy and the well-being of Soviet citizens.

George Heikens and Guy Bush had similar agricultural backgrounds. The two men grew up on mixed-use farms in Iowa, and both delayed attending college in order to work on their families' farms. In college, they had both studied hog breeding, the skill that captured the interest of their Soviet employers. When they signed their contracts Heikens was just about to graduate from Iowa State College at age thirty, and Bush, at thirty-nine, had worked for two years as a writer for *Wallaces Farmer,* the Midwest's most popular agricultural periodical. By the spring of 1930 the Great Depression had dimmed work prospects for both men, and they considered themselves lucky to be among the animal specialists hired to work abroad. The Soviet Swine Trust would pay them over one thousand dollars for their year abroad, a far more generous salary than they were likely to earn in Iowa. The trust would also cover travel and living expenses. And so, in the blistering heat of late July, Heikens and Bush packed trunks full of the warmest clothes they owned and set off separately, first by train, then by Atlantic steamer, for a place they both called Red Russia. They had almost no notion of what awaited them across the ocean.[3]

Heikens and Bush were hardly alone in their lack of knowledge. In 1930, most Americans had little idea how farm collectivization or other aspects of socialist rule had changed the everyday lives of rural citizens in the Soviet Union. During their terms of service Bush and Heikens would experience the disorganization and resentment of a typical Soviet collective farm firsthand, without understanding the historical conditions that had led up to the dramatic decision to attempt to collectivize and modernize Soviet farms as rapidly as possible. However, it is important to be familiar with the recent history of the Soviet countryside in order to understand what the state was trying to accomplish, as well as why the task proved to be so challenging.

In the eyes of the Soviet Union's leaders, Russian agriculture before collectivization "resembled a boundless ocean of small individual peasant farms with backward, mediaeval technical equipment" in desperate need of reform and repair. In spite of this need, the Communists who controlled the Soviet Union adopted a policy of benign neglect toward the countryside during their first decade in power, a situation with which most rural residents were completely content. The earliest agricultural legislation the Soviets passed made no mention of collectivization; in fact, in 1921 the new Soviet authority promised to reduce the tax burden for "industrious peasants who increase[d] the sown-area and the number of cattle in their holdings and . . . the general productivity of their holdings." In other words, the state would reward capitalist behavior that increased farm productivity and the amount of food available for the market. In this early era, Bolsheviks focused first on establishing socialism in the cities; building socialism in the countryside was a secondary concern.[4]

Left to their own devices, farmers built up their stocks of animals, invested in new tools and seeds, and improved their barns and houses during the New Economic Policy, or NEP, period, which encompassed most of the 1920s. Farmers paid lower taxes and received more cash for their crops during NEP than in any other period of Soviet rule. Popular with producers, the high food prices of the NEP era were unpopular with consumers in cities. By the late 1920s many Bolsheviks, including Stalin, supported universal farm collectivization in order to give the state more control over grain prices. Until collectivization, farmers acted as their own agents and had considerable influence over the price of grain crops. They could withhold grain from the market for several months in order to drive prices up, thus significantly improving their profit margin. As bread prices rose in turn, an increasing number of Soviet officials became convinced that the state needed to intervene in order to curb the greed of farmers and regulate the supply of all grains, especially wheat. In the winter of 1929 the NEP era of relative prosperity and calm ended with mass farm collectivization drives.[5]

Although these campaigns collectivized over 80 percent of Soviet farms in just a few months, the drive was not the first time a zeal for collectivization had struck Soviet bureaucrats. A collectivized farming system had always been the ultimate goal of Soviet rural policy. Throughout the 1920s, the government had debated and experimented with various forms of collaborative agriculture—creating trial communes (artels) and collective farms of different sizes and with various types of cooperative organization—but administrative disorganization, scarce machinery, and a lack of clear direction from Moscow kept the countryside relatively free from heavy-handed state interventions until the end of the decade. No single form of state-sponsored agriculture became common on a national scale during the 1920s. When the Central Committee first debated creating a national policy of collectivization in 1926, early arguments focused on whether rapid collectivization was even possible. Nikolai Bukharin, the principal architect of NEP, argued that collectivization should take years, if not decades. Joseph Stalin, however, believed that rapid collectivization was necessary in order for agricultural industrialization to occur, and that a "whirlwind" approach was best.[6]

The whirlwind began in November of 1929 when the Central Committee voted to modify the first Five Year Plan in order to include farms and rural areas. The reforms spelled out in the revised Five Year Plan consolidated and collectivized the majority of Soviet farms. State planners grossly underestimated how unpopular this sudden move to collectivize would be in the villages. Both before and after NEP, Soviet farmers thought of themselves as freeholders and the descendants of recently emancipated slaves, entitled to the land they worked as a condition of emancipation. Many believed state collectivization would be slavery by another name. Indeed, mass collectivization was usually accomplished by force or coercion and kolkhoz workers were rarely paid in cash. Beginning in late 1929, the state rapidly consolidated private fields, confiscated farm animals and field tools, and persecuted those who refused to join the collectives.

Collectivization began as an orderly if ambitious campaign of property mergers, but it swiftly devolved into state-sponsored terrorism. During the first months of the campaign, the state offered small cash rewards to farmers who joined the kolkhozes. This was largely unsuccessful. When bribes and propaganda did not convince peasants to collectivize, the state turned to violence, arresting those who resisted and forcefully confiscating farm tools and animals. During this period, the state also intensified its campaign of dekulakization, which identified and persecuted capitalist farmers in agricultural communities. Since farmers had been encouraged to make capital investments in their land, animals, and tools throughout the NEP era of the 1920s, every village had numerous potential kulaks, but identifying and vilifying successful farmers was a subjective process. Inevitably, villages that had most forcefully resisted collectivization were found to contain the most kulaks. The process of dekulakization killed some peasants, caused others to be arrested, disenfranchised entire families, and banished them to distant provinces.[7] Perhaps the most effective result of these campaigns was to produce immediate and severe consequences for communities that protested collectivization, turning popular sentiment against collective protest and toward communal obedience to the state.

Peasants resisted collectivization most forcefully during February and March 1930. To protest the state's whirlwind campaign, rural residents organized marches, occasionally attacked regional officials, and slaughtered horses, pigs, and cattle that were slated for collectivization.[8] In some regions protesters killed a quarter of all livestock. These protests did not reverse or completely halt collectivization, but they curbed its early whirlwind. In its place, a two-year détente developed between the state and the countryside and the government replaced force with inducements. Villages that collectivized willingly received extra money, better seeds, more animals, and more technical support as rewards. Many of these compliant farms became model kolkhozes, which were intended to become showcases of functional collectivization. Extra assistance from the state was definitely an advantage, but these advantages did not necessarily guarantee the successful modernization of model farms.

Historians have often focused on the problems Soviet collectivization created for grain production and distribution. Grain, especially wheat, was a staple of both the Soviet economy and the Soviet diet, and ensuring a reliable and affordable supply of it was a central task of the government. However, almost 10 percent of newly collectivized farms specialized in raising animals, not grain, and the state paid special attention to these farms after mass protests resulted in the slaughter of so many livestock. The state devoted extra and disproportionate attention to kolkhozes that specialized in meat or milk because they were early proving grounds for the government's ambitious plans of agrarian modernization. Before collectivization, a minority of farming operations specialized in raising livestock and there were not many Soviet agricultural professionals trained in feeding, breeding, or veterinary care of farm animals. The People's Commissariat of Agriculture struggled to fill these gaps in expertise, often deploying veterinary field technicians (known as vet-feldshers) to new collective farms and, in 1930, hiring American experts to help organize new specialty kolkhozes.

Creating farms that were dedicated to producing just one item was a new tactic that the Soviet state adopted for both grain farms and livestock operations. Although most kolkhozes remained multipurpose farms for another two decades, a few were selected at the very beginning of collectivization to become models of a new, more modern style of specialized agriculture. Specialization was intended to make collective farms more efficient. Communist officials, kolkhoz managers, and other professionals found the contemporary system of farming in the Soviet Union to be appallingly inefficient and that this was the central flaw of Soviet agriculture. If farms could be made to be more efficient, either by scaling up and completely mechanizing, or by specializing in just one marketable product, then their profitability would also increase. The Soviet state experimented with both tactics, creating massive, highly mechanized grain farms on the eastern frontier, and closer to urban centers, establishing model farms that produced high-value products like meat and milk. Neither tactic worked well. Moshe Lewin has noted that "the first [Five Year] plan . . . produced a kind of self-perpetuating mechanism in which uncoordinated and quite arbitrary economic targets served to enlarge the scope of 'planning' without necessarily improving order or efficiency on the ground."[9] Inefficiency, both real and imagined, plagued the agricultural goals of the state for the next two decades.

Traditionally, in Russia, the marginal nature of most agriculture made crop diversity a necessary adaptation. It also encouraged self-sufficiency. While regions specialized in cash crops such as flax and cotton before collectivization, almost every farm also strived for a basic level of household autonomy, raising vegetables, feed crops, and root crops for personal and animal consumption. After collectivization, the Soviet state decided (often in an arbitrary manner) which farms would specialize in specific crops, and which farms would continue producing a more diverse range of products. Both of the farms highlighted here—Rodomanovo in Western Oblast (roughly present day Smolensk Oblast), and Millerovo in North Caucasus Krai (roughly present day Rostov Oblast)—were selected to specialize in raising pigs. Before collectivization, farmers in Rodomanovo had raised flax, dairy cows, clover, and rye (and almost no pigs). In Millerovo, farmers had grown wheat and sunflowers and raised pigs, in addition to maintaining large vegetable gardens.

It was to further the conversion of these farms that the Swine Trust offered contracts to at least three men, including George Heikens and Guy Bush. They were charged with "working out and applying of all manner of work in the sphere of pig husbandry, that is the mechanization of hog house equipment, [and] the application of more perfect means of feeding, breeding, growing and fattening of pigs."[10] Experts such as Heikens and Bush were hired to introduce an American style of farming expertise to Soviet model farms, but this proved to be an

Map 2. Western Oblast in 1930

Map 3. Present-day Rostov Oblast

impossible goal. In the first decades of the twentieth century, American farms had mechanized, scaled up, begun to specialize in single crops or single species of livestock, deskilled farm labor, and dramatically increased the number of university-educated agricultural professionals. American farming in 1930 depended on machines, price supports, and good roads and rail lines. Livestock operations had incorporated intensive feeding regimens, and advances in the nascent science of genetics, such as artificial insemination and mathematical formulas that predicted the level of inbreeding in related pure-bred animals, had revolutionized animal reproduction. None of these elements existed in the Soviet Union, and all of them proved difficult to introduce.

Bush and Heikens are potentially problematic eyewitnesses in this story. They spoke no Russian and both of them relied on interpreters during their stay in the Soviet Union. Neither had ever traveled outside of the United States before this trip, and both men, while curious about the Soviet Union's experiments with communism, regarded Soviet policies with bemusement and, occasionally, disdain. The histories set out in their letters reveal as much about their own prejudices as they do about the everyday activities of the villages where they lived. But their letters also offer candid and unique critiques of the new collective farms of the Soviet Union. In the case of Heikens, archival evidence supports many of his observations about the disorganization, poor leadership, and widespread dissatisfaction among foreign workers he witnessed on his farm located just a few hundred miles from Moscow. Contemporary local archives for the oblast where Guy Bush was stationed were destroyed during the Second World War, but accounts from travelers to the region support many of his observations about the lack of food, the burden of dealing with

irresponsible chairmen, and the everyday hardships that kolkhoz workers in the grain belt of southern Russia and western
Ukraine faced.

It is remarkable how often the letters of Bush and Heikens echo the frustrated attitudes toward kolkhoz workers that
Soviet officials held during this period. In the words of Merle
Fainsod, who worked closely with the collectivization-era Party
archives of Western Oblast, "The general impression derived
from the archive is one of apathy among the rank and file and
a notable lack of enthusiasm for life in the kolkhoz. . . . [T]he
Archive is full of complaints of kolkhoz inefficiency, of lax
work practices, of drunkenness, of thievery and of even worse
abuses."[11] This bleak description of rural daily life matches the
observations Guy Bush and George Heikens made in their
respective villages after just a few months of work in the Soviet Union. The increasing pessimism of the two Americans
might be attributed to their cultural isolation or the chronic
rhetoric of totalitarian power, but it seems far more likely that
Soviet and American experts shared a mindset when they approached the work of modernization.

This "will to improve" is a widespread phenomenon in
agricultural modernization and industrialization, and it allowed both groups to overlook the agency of kolkhoz workers upon whose labor their projects depended. Both sets of
experts treated rank-and-file workers, almost universally, as
an untrustworthy liability. It is also, of course, entirely prob-
able that the work environment of a Soviet kolkhoz in 1930
was frustrating and inefficient, with few managers or workers
showing the zeal for progress and development that so many
Americans had noted in urban managers and workers just a
few years earlier.[12]

Frustrations with personnel allowed American and Soviet experts to overlook the environment as a major impediment to the success of farm modernization. Heikens and Bush wrote constantly about the weather, animal diseases, mud, and poor winter feeding regimens, but they very rarely blamed the natural environments of their respective farms for their difficulties raising pigs. They consistently noted when incompetent managers or untrustworthy workers hampered their efforts, but animals often died and projects failed primarily because of the challenging natural environment. Poor management and poor work organization exacerbated biological limits to growth but they were not the principal causes of high mortality on either farm: these were disease and cold weather.

While definitions of industrialization typically focus on labor organization, mechanization, and the scale of production, the experience of the first generation of collective farms in the Soviet Union exposed a necessary category of industrialization that was initially ignored by Soviet managers to devastating effect: that of anthropogenic control over nature. Gaining control over the natural features of animals and plants on farms proved more difficult and more expensive for the Soviet state than gaining control over those of waterways or forests. Farm animals in particular were hard to keep healthy and alive; good management was simply not enough to improve their lot in the same way that professional managers with strong visions influenced forest conservation or canal construction during the same era.[13] Soviet managers were not in a position to seize control of nature even on well-funded model kolkhozes in 1930. As a result, collective farms foundered during the 1930s, limiting the productive capacity of agriculture across the Soviet Union, as well as future prospects of a Soviet, socialist model of industrial agriculture. The American experts the Soviet Swine Trust hired to help manage new farms witnessed this failure.

George Heikens, the younger of the two, was assigned to live and work in the village of Rodomanovo, 150 miles northeast of Smolensk. In 1934, Yuri Gagarin would be born just eight miles from Heikens's farm, and today the district is named in his honor. In 1930, the village of Rodomanovo was in the Gzhatsk District of Western Oblast on a former nobleman's estate. The region had historically been poor—peat bogs and pine forests created acidic soils that made farming difficult —but farmers across Western Oblast had improved their fortunes during the last half of the nineteenth century by raising flax, a booming export crop that was eagerly purchased by Europe's rapidly expanding navies. In 1900, 20 percent of the arable land in the oblast was planted with flax. Other important crops included potatoes, oats, rye, and clover. Farms near rail lines had founded dairies and taken to pig breeding in the early decades of the twentieth century, but in Rodomanovo's district, only 1,467 pigs were listed in a farm inventory in 1930, which makes it unlikely that any commercial pig farms had existed near Rodomanovo before the kolkhoz was tapped to specialize in pigs. Village crafts and out-migration to larger cities were important secondary sources of income for the residents of Western Oblast. Except for a handful of textile mills, the region had almost no industry. Like most of the Soviet Union in 1930, Western Oblast was predominately agrarian. In 1930, 82 percent of the population lived and worked on farms. In 1930 on kolkhozes in Gzhatsk, Rodomanovo's district, 85 percent of the 5,442 members were listed as having either "poor" or "average" incomes. Only 371 members, or 7 percent, were classified as "prosperous." NEP may have been a time of stability in Gzhatsk, but it had not created a robust class of wealthy rural elites.[14]

Once it was slated to become a model pig farm, Rodomanovo needed a swine expert, and Heikens fit the bill. He wrote proudly to his parents in September 1930, "Our farm is to be a model one and we are to have anything we want to make it so."[15] When Heikens arrived most of the pigs he saw on-site were small, feral specimens, but the Swine Trust assured him it that had already ordered new purebred sows from Germany that would farrow in October. For his first few weeks on Rodomanovo, Heikens's main concern was preparing living quarters for the new pigs and, once they arrived, keeping the sows and their piglets alive. His letters home rarely mention the workers he supervised during these early months, except to remark upon their astounding inability to perform even the simplest task around the pig barns correctly.

Compared to George Heikens, Guy Bush was a more experienced traveler and a more seasoned swine expert, but he also found himself in unfamiliar surroundings once he arrived at his assigned posting, Kolkhoz 22, forty miles outside the village of Millerovo in the Rostov region of southern Russia. Millerovo was a wheat-growing community located at the edge of the Black Earth Region. Since the late nineteenth century, it had also been a center of production for sunflowers, which were harvested for their seeds, their oil, and as a winter feed for livestock. Although Kolkhoz 22 was supposed to be a cutting-edge, hundred-thousand-acre enterprise, it still felt like a sleepy, old-fashioned southern village to Guy Bush when he arrived in September 1930. Bush noted that its church bells tolled every morning and that the kolkhoz members were skilled craftsmen who sewed their own clothing and thatched their tiny two-room cottages with straw. On warm evenings, the people gathered in the village square to sing folk songs and dance, and many villagers retained the habit of constantly nibbling sunflower seeds, spitting the hulls indiscriminately on the floors of houses, trains, or theaters.[16]

Although their material living conditions were quite different, Guy Bush and George Heikens responded similarly to the work situations they encountered in the Soviet Union. On Heikens's farm there was plenty of housing available for workers and families. The food he was served was bland, but there was always enough of it. Although outsiders appointed by a regional Communist Party committee ran most collective farms, the manager of Heikens's farm, Aleksandr Kotov, was an agricultural professional from the local community. In contrast, the living conditions on Bush's farm were more austere. Housing was competitive, and he was isolated, located forty miles from the nearest town with a rail connection. Bush also judged that he had the worst food situation among the Americans he knew living in the Soviet Union.

While regional differences mattered, and Rodomanovo was calmer and more prosperous than Kolkhoz 22, both men experienced some of the hardships of collectivization firsthand, including the winter of 1931, which was especially long and snowy. When Bush's employer asked if he would consider remaining on Kolkhoz 22 for another year, "I could only answer that I could not continue under the present living and working conditions. Engineers are treated quite nicely but the agricultural force is looked upon as—(I hardly know what)."[17] Although both men had access to the best local foodstuffs available—in the fall, Bush was the only member of his village who was allowed to purchase a sheep carcass—both complained of its monotony, especially the lack of fruit and garden produce. Heikens mentioned in August of 1930 that "the food here is good, but not as good as it was on the boat [over]. Have had rice with milk several times lately." Bush noted that many could hardly be expected to do much hard manual labor over the winter, because "black bread and thin soup is a poor cold-weather diet . . . the families all [live] in a two-room house [with] little to cook and with conditions as they are, apparently some have lost hope." Food and housing shortages in the North Caucasus Krai intensified over the next two years; Bush's hard experiences foreshadowed the deadly famine that would hit the region especially hard in 1933.[18]

Aside from the food, however, the first few months in the Soviet Union were a busy time for both men. Inexperience and disorganization at Rodomanovo were the first problems Heikens mentioned in his letters home, but this observation did not initially erode the attitude of goodwill he brought with him to the Soviet Union; that came later. His early letters reveal an adaptable and even-tempered man intent on doing the job for which he was hired as efficiently as possible. For example, upon arriving at Rodomanovo, he wrote that he was disappointed to learn that the farm had forty sows due to farrow in the frigid months of December and January, a dangerous proposition on even a well-established farm. However, Heikens did not complain about such poor planning but instead pushed to install central heating in at least one building before the piglets' arrival. Bush also immersed himself in the work of his farm. By mid-September he wrote home to his wife, "I am not quite so lonesome as I was. There is too much to think about and I feel quite responsible in doing the best I can."[19]

Both men shared an initial enthusiasm for both the work and their co-workers, and for most of the fall both Bush and Heikens remained optimistic about their ability to effect change on their respective farms. When he first arrived in Moscow

Heikens wrote home, "There is an unlimited opportunity for me to help here. They need our experience very much and certainly need more livestock." On his first day at Rodomanovo he opined, "The managers do not know much about hogs if you ask me."[20] After one month on the job, the director asked Heikens to make recommendations on how to improve the farm. His suggestions were wide-ranging but essentially called for a complete overhaul of operations in order to bring them in line with what he knew would work on an Iowa farm. Food, housing, and general animal management were substandard. Pigs should receive dry rations with less fiber, not the wet steamed potatoes they were currently being fed daily. Adult pigs needed rings through their noses to curb rooting, and they should be let out onto fenced rather than open pastures, which would allow the farm to dismiss the twenty-four (in Heikens's mind, superfluous) swineherds it employed. New piglets should not have their milk teeth cut out, which led to infections, they should be castrated and weaned earlier, and their ears should either be notched or tattooed with a unique identifying number at five days. Every building at Rodomanovo needed a new wooden, sloped floor to replace the earth floors of the barns and outbuildings, the roofs leaked and needed repairs, and the new ventilation system was not effective.[21]

If Guy Bush wrote similar letters of advice to his superiors at Kolkhoz 22, they no longer exist, but he did outline a week's worth of work in September in one of his letters home. Many of his activities overlapped Heikens's. In one week Bush held two conferences with the farm manager and another with the manager of the hog department. He wrote out an order for materials with which to build pigsties, and one for fish meal and cod liver oil supplements. He inspected a litter of dead pigs and a shipment of 571 pregnant sows. He wrote a report that offered suggestions for improving rations on the farm, visited (on horseback) seven herds of hogs that were quarantined at a distant location, wrote up a proposal that all hog house construction to be made of wood, and drafted a plan to prevent piglet deaths in cold weather. Finally, he assisted at the autopsy of a boar that had died from cholera the week before.[22]

The efforts Bush and Heikens made to improve their assigned farms challenged numerous established policies and routines. As outsiders in the Soviet Union, neither American fully understood the costs and hardships that might be associated with the improvements they initiated. The early suggestions Heikens made strained both the budget and the abilities of Rodomanovo's staff, and Bush's orders of feed supplements and building materials never appeared, presumably because such products were either too scarce or too expensive. Nevertheless, the central office of the Swine Trust theoretically supported complete farm overhauls in order to specialize in the large-scale production of pigs. Initially, Heikens's many requests were honored. In October he wrote, "They are building [sty-separating] gates to my drawings. Make them exactly like the drawings and even noticed where I had not drawn in the nails on several boards. I believe gates will be lots better than the bars they have now." On Kolkhoz 22, Guy Bush's suggestions were also taken seriously. In October, after a meeting with the chief inspector, the veterinarian, the kolkhoz chairman, and the executive bookkeeper, Bush wrote home of the meeting that "It was necessary to tell them many unpleasant things about the way they were doing things. . . . We opened a bottle of 'cognac' (a French whiskey), and I believe we are better-acquainted and better friends now. It was an ordeal that I dreaded to go through (telling the facts), but that's what I'm here for."[23]

The two men began to lose their optimism around the same time, in December 1930. One of the first things they both complained about in their letters home was the poor leadership on their respective farms. This was a common grievance across the Soviet Union in the first years after collectivization. Each farm had a chairman who was typically appointed from outside the community. Some chairmen were lazy or too easygoing, but most were "hard driving, hard drinking, blustering and threatening, frequently abusive and foul of mouth."[24] In the first years after collectivization, the vast majority of kolkhoz chairmen were unpopular with workers, experts, and bureaucrats alike. Identifying competent, sober workers who were literate enough to fulfill the considerable bureaucratic demands of the position of chairman was an ongoing struggle for the regime. Regional Party committees appointed chairmen, and because this was an appointed position, kolkhoz members could not simply vote them out if they were unpopular. If a kolkhoz met its production quotas, it was almost impossible to remove a bad chairman.

Merle Fainsod has noted that during this early period of collectivization, kolkhoz chairmen often strategically sought to ingratiate themselves with a few people in powerful positions.[25] Although American experts did not hold traditionally understood positions of power, Heikens and Bush initially noted that the chairmen of their farms befriended them. The manager of Kolkhoz 22 tried to give Guy Bush a porcupine, "but he was too dirty to keep in the room [and] every time I tried to pet him he handed me a quill . . . [but] they want to make me satisfied. A big fur lined vest has been given to me; now they are making special boots." Heikens was friendly with his farm director, Kotov, as well. The two seemed to have a good rapport during the fall of 1930; Kotov ordered a hunting rifle from the Sears Roebuck catalog Heikens brought with him, and the two men went out for Sunday walks together throughout the autumn.[26] The general rural population may not have been in a position to curry favor with foreigners, but they were certainly very interested in the American visitors to their villages. Bush and Heikens both noted they were often treated as celebrities in their daily lives in the rural Soviet Union. Strangers sacrificed their seats in theaters to make sure the men had a good view of the stage. Train passengers gave up their sleeping bunks in railway cars so that the Americans might rest on overnight trips. In rural areas the American guests were objects of great fascination; Heikens noted on his first day in Rodomanovo, "I'm like a God around here." Bush, visiting a local fair in the town of Kashary, wrote, "The crowd made it almost unbearable, for they crowded around me until there was hardly breathing room. . . . I have penetrated a community where foreigners are scarce so I am still a novelty." For the most part, foreign visitors to the Soviet Union remained in the major cities or worked in newly created centers of manufacturing such as Magnitogorsk, a steel-producing town, or Stalingrad, where tractors were built. Although a few newspaper reporters and intrepid tourists had made brief visits to the Soviet countryside, Heikens and Bush were two of the very first Americans to live and work in the rural Soviet Union.[27]

Bush and Heikens also grew frustrated with the amount of work they were expected to accomplish. Heikens wrote to his parents, "I have to do lots of little things which cheap help could just as well do if they knew how," and "the workmen here know absolutely nothing about hogs, they can't drive one 10 feet. . . . If I wasn't so busy, this would be a lonely place. There are no movies, nothing to read and no one to talk to." Once Heikens was familiar with the organization of the farm,

his complaints sometimes adopted the tone of a Soviet bureau-
crat: "The place is in a state of disorganization all the time," he
wrote in December.[28]

Guy Bush noted he was too busy to write letters to friends
or colleagues back home because he had "only one worker who
knows anything about swine. The two of us must train help that
is somewhat slow in grasping details to care for many thousand
hogs." He often complained about the lack of modern feeding
practices on his farm; the Swine Trust initially promised him
one hundred tons of soybeans for his new hogs, but when he
went to pick up the feed, he was offered only four and a half
tons of inedible woody stems from soybean plants. "It is an
illustration of how little they know about their own organization," Bush wrote. "I am doing considerable manual labor, not
that it is necessary, but I find it is the only way to train the
workers. They do not know anything about the minor details
in caring for swine. There is a terrible waste of labor, which
must be eliminated." This same observation appeared in the
archive of Western Oblast for the entire span of the 1930s.
Fainsod notes that "The OGPU reports from the raions strike
an almost monotonously repetitive note of total disorganization and inefficiency."[29] Especially noteworthy in the descriptions of both Bush and Heikens is how closely their words
echo the stereotypical criticisms of Soviet leaders at this time
as they railed against ignorance, apathy, and inefficiency on
the collective farms.

In spite of these strains, neither Heikens nor Bush conflicted directly with their managers until the end of 1930, just
as winter descended. In December on Rodomanovo the fodder and bedding supply deteriorated dramatically, pilfered by
workers who had little access to these necessities for their privately owned livestock. At the same time, Rodomanovo's pig

population grew due to both births and new arrivals. The new animals added stress to operations. By December a population of three hundred pigs had multiplied into twelve hundred, and Rodomanovo's staff was clearly overwhelmed by the additional animals. It was not just the sudden population explosion on the farm, but also the vulnerability of pigs to disease, overcrowding, and cold that stymied progress. Yet Heikens continued to single out human errors and poor management decisions as the central problems on his farm. For example, in early January, against Heikens's orders, workers fed a group of pregnant sows spoiled, alcoholic malt that caused nine of them to abort their litters and another five to farrow a week early, resulting in the deaths of almost all of their offspring.[30] Heikens had noted earlier in the fall that farrowing sows and newborn piglets would be a major liability in the harsh winter months of December and January, but when these particular litters died en masse, as he had predicted they would, he blamed the mistakes of the caretakers rather than the weather.

His attitude toward the farm's workers did not improve over the course of the winter. In February, he wrote to his parents, "Don't know if [evening classes he taught] help much or not. Sometimes I think the workmen will never understand hogs. They throw in some feed and walk off." In the spring he expressed the same sentiment: "I have to watch the workmen pretty closely, they sometimes forget to feed a whole group of hogs and never put enough straw in the pews if no one is watching," and "every fourth day" he complained, he had to explain their jobs again, "as they seemed to forget the routines that quickly." Poor work habits might have stemmed from low morale in the aftermath of collectivization. Forgetting daily routines and other foot-dragging activities may have been one of the only relatively safe outlets Soviet farmers had through which to express their dissatisfaction with the recent changes they had experienced on their farms.[31]

Both Heikens and Bush were aware of political purges that had happened during the era of collectivization, and made allusions to the unrest and violence that had preceded their arrival. Although they expressed distaste for the harsh punishments bad managers received, neither of them refrained from calling out incompetent managers, even though they knew that convicted managers might be imprisoned or even killed. Bush wrote at the end of November, "I sit here like a little tin god, removing those from office who fail to produce. Often a life is at stake. I can't say that I like it—but there is no pleasure as an American in being connected with a failure so I've started to wade ruthlessly through them to make a showing while I'm here." By early April, however, Heikens formally complained about Kotov to the Swine Trust: "I explained to Mr. Kotoff many times . . . that I am very particular about the system of feeding sows at farrowing . . . and he saw it in use at [another collective farm] but disregarded it and did not tell the workmen to feed differently. . . . Results: a greater chance of having scoured pigs." Heikens's complaint about his manager deflected some of the blame for a failed project from himself to another. Heikens had written home earlier about the mistakes of workmen who misfed their charges, but in this letter, written two months later, it is Kotov, not the workers, whom Heikens holds responsible for poor feeding practices. Kotov had failed to "tell the workmen to feed differently," and it was this lack of managerial authority that had resulted in illness and death. Heikens left Rodomanovo in May 1931, and Aleksandr Kotov's fate does not appear in the regional archive records. It is unknown whether or not Kotov was able to keep his job, but during these years the turnover of farm management was high, with many deposed chairmen charged with criminal mismanagement.[32]

Bush and Heikens were not unique in finding the Soviet culture of denouncing incompetent bosses normal. David Engerman has noted that British and American educators who toured Leningrad and Moscow during the early 1930s accepted such denunciations as a matter of course.[33] Heikens and most of his peers were not communist sympathizers, but they believed the Soviet Union would be better off as a developed and industrialized nation. In participating in Soviet agricultural modernization, they also felt obliged to participate in the policy of recrimination that characterized Stalinist management during this period. Indeed, it may have been difficult to avoid condemning co-workers. Although Heikens's letter complaining about Kotov appears to be spontaneous, Heikens and Bush were urged to identify the worst managers on their farms and corroborate accounts of incompetence that others had submitted to the Swine Trust. Their letters reveal that privately, Heikens and Bush found many workers and managers incompetent and unskilled. Speaking out against a colleague could have dire consequences, but Bush and Heikens were reluctant to remain silent because they risked being blamed for problems for which they were not responsible. Bush even brought a sense of patriotic duty to the task when he noted, "there is no pleasure as an American in being connected with a failure."

In April, George Heikens found himself on the receiving end of accusations of mismanagement. After the high death rate of piglets in the winter on Rodomanovo, a wet spring brought even more of a die-off to the herds of pigs that he managed, which also meant increased surveillance by outside engineers and officials. He wrote to his parents in April, "we have visi-

tors almost every day. Some like the looks of things and some
don't. I'm getting so I pay little attention to what they say because you can't please all of them and lots of them never raised
a pig in their lives." Although Heikens never feared for his life
or liberty, he still found these inspections and the suspicion
they implied both aggravating and unfair. Heikens wrote to the
Swine Trust, "I feel that regular inspection is necessary, but . . .
I do not like to have people tell me that ours is the worst farm
in Russia and that the imported sows are doing fine on other
places when I know of several farms where not all is smooth
sailing." Although none of the many visitors to Heikens's farm
in the spring of 1931 blamed him directly for Rodomanovo's
failure to meet the high expectations for production set by
Moscow, the sheer volume of outside evaluations the farm
received indicated that officials were concerned, and these
visitors were not shy about voicing their opinions about the
operation. Heikens wrote to his supervisors, "We had visits
from very many engineers, journalists, veterinarians and general inspectors, all of whom criticized freely." The letter Heikens wrote to the Swine Trust about Kotov's shortcomings was
penned in the days immediately after he had received a flurry
of critical official visits. His critique of his manager could easily be interpreted as a defense against criticism that had originally been directed at him.[34]

The desire of Bush and Heikens to see their projects succeed trumped more accepted social obligations to safeguard
the life, liberty, and job security of their co-workers. This per-
version of professional values was a hallmark of early Stalinist
project management. Rather than create and maintain a cohort
of competent professionals who worked together on a single
project from start to finish, Soviet projects of the early 1930s
typically assembled a group of people of varying skill, and ex-

pected that long work hours and a zealous commitment to the task at hand would compensate for a lack of professional training. This was not simply carelessness on the part of the government; there was also a shortage of skilled workers. Importing American, British, and Western European professionals was a partial solution to this shortfall; another partial solution was to simply appoint Soviet workers to jobs for which they were not qualified, hoping that natural aptitude and the threat of dire consequences in the face of failure would inspire them to rise to the challenge of the task. While it was difficult to remove managers whose farms performed well, periodic purges swept administrations from unsuccessful farms. This approach was terrible for group cohesion but was excellent at keeping workers motivated to work long hours for low pay. In the absence of good organization and competent cadres, violence and the threat of violence maintained order and kept complicated projects on track.

Bush, Heikens, and Soviet government officials all attributed the main problems of animal farms to human error and a lack of rural infrastructure. Theories of modernization called for improving basic infrastructures in order to make them look like those in wealthier, more industrialized countries, presumably creating a more efficient working landscape. An industrial agricultural infrastructure that included reliable roads, tractors, scientific breeding programs, and standard bookkeeping practices was entirely absent in the Soviet context immediately after collectivization. Certainly some of the problems that foreign experts and Soviet officials encountered on kolkhozes were due to these deficiencies. The archives of Western Oblast as well as the letters of Bush and Heikens describe numerous instances of disorganized workers whose incompetence threatened to derail the work and success of collective farms during this critical period. While the Soviet state and professionals like Heikens and Bush focused their attention on faulty humans, worker error was not the only factor that threatened early collective farms.

The most insurmountable barrier to a thriving pig farm at both Rodomanovo and Farm 22 was natural, not cultural. Both places were cold and disease-prone, and food was scarce. In an era before antibiotics, diseases moved through herds rapidly, especially herds kept in close indoor quarters all winter. The state exacerbated some natural limits by scheduling hundreds of piglets to be born during the coldest months of the year and not planning winter food rations far enough in advance, but the biggest challenges for Soviet pigs were environmental, not managerial or organizational. Animals and diseases played just as significant role in the struggle of early collective farms as the flawed workers to whom managers paid so much more attention.

Pigs, the animals that Heikens and Bush tended, are potentially excellent industrial animals because they respond well to minor changes in their diet, housing, and hygiene. For example, increasing the protein in a pig's diet by just a few percentage points corresponds with a dramatically improved feed-to-weight ratio. Although pigs are naturally sickly—prone to cholera, scours, pneumonia, and bacterial infections—they are relatively easy to keep alive if kept indoors and away from cold air for the first few weeks of their lives. However, while modifications like better food and improved housing sound straightforward and low-cost, they proved nearly impossible for first-generation Soviet collective pig farmers to implement, even with the help of overconfident American experts like Heikens and Bush.

For example, the ten villages that made up Bush's model

collective farm, one of the largest in the whole Soviet Union, was intended to house twenty thousand hogs as well as a thousand head of cattle. It is unclear whether anyone running either Bush's or Heikens's farm understood enough about raising animals to recognize that industrial farm operations could not rely on grass and hay to feed such a large number of animals. At one point, Bush's manager suggested that new animals be fed straw (more typically used for bedding than for feed) to make up for shortcomings in the diet. Bush noted early in his stay that "farm feeds vary a great deal . . . and they have not as yet appreciated the value of some of the necessary feeds to properly produce swine." Later he complained, "I have no proteins like milk, meat meal or fish meal to work with. There is no alfalfa either."[35] Bush himself, as the most privileged eater on the farm, had only sporadic access to milk and meat; the Soviet Union had none of these products to spare to feed the animals it planned to raise.

Farm managers struggled with an appropriate response to such challenges, often overreacting or responding in ineffective ways. Before pigs ever arrived on his farm, Guy Bush recognized that the sheds that would house them were placed too close together and disease would spread quickly among infected populations. As Bush predicted, the pigs fell ill in the early winter, but Soviet authorities were intent to single out a scapegoat. On Bush's farm, blame fell on "a poor old veterinarian who I had esteemed up to now, has been one of the few *honest people* with whom I work. . . . he is accused of hindering the development of a government project. If the accusations are true, he will pay with his life. . . . the present system here has developed some crafty 'blamers.' The better they can blame or transfer responsibility to the other fellow when things go wrong the better job they hold."[36]

Housing for animals became another point of contention on both farms, and again, these struggles point toward a rigid set of requirements built into the nature of pigs that the Soviet material reality in 1930 simply could not meet. On both farms, planners assumed that they could easily construct entirely new outbuildings that would be modern, hygienic, and able to accommodate the huge influx of new animals. In all instances they were mistaken. On Bush's farm, animals were supposed to be kept indoors all winter. This had originally been the plan on Heikens's farm as well, but after his protest, this plan was changed. In part, exercising pigs during the long winter kept their strength up. It also exposed them to sunlight, which helped them manufacture vitamin D, the lack of which damaged their immune systems. The houses were too drafty on Heikens's farm, with no glass in the windows to shelter younger pigs from the brutal northern winter or allow sunlight in.[37] In both cases, regional authorities provided materials and supplies for constructing new buildings while overlooking important details —such as glass for window panes—that would later make the difference between success and failure.

Fences and other enclosures were other basic items of animal industrialization that early farms lacked. Fences were not common in the Soviet Union of the 1930s, either before or after collectivization, and planners initially overlooked the need to build enclosed pens, since there were plenty of workers who could work as swineherds. Fences were seen as tools of capitalism, delineating the boundaries of privately owned property. However, while human swineherds are good at driving pigs and keeping them away from predators, they are useless at preventing diseases from spreading from pig to pig or from passing into pigs from the ground. Industrial hog production isolates swine populations not simply to limit procreation but also to prevent the spread of disease. Heikens's main complaint about the farm at Rodomanovo was that "they are putting too many hogs together in large herds to suit me. I believe they are heading for trouble in their hog farms. . . . They won't take our advice at all when it is against a plan of the government, they just figure the government is always right."[38]

Bush complained to his wife and family about the lack of good management on his farm, but it was the health of the pigs and their untimely deaths that delivered the most crushing defeats to his work. Diseases were varied, acute, and seemingly insurmountable. At his arrival, Bush had noted that cholera was endemic to the herd.[39] During the cold winter, the smallest pigs succumbed to pneumonia and "scours" or serious diarrhea. In the early spring a feed regimen that included few vitamins meant that growing pigs became severely malnourished and got rickets, a disease caused by a lack of vitamin D. Heikens had the same experience on Rodomanovo; the principal killers of pigs were endemic diseases like scours and cholera, not human managers. While poor management and incompetent workers sometimes made illnesses worse, both farms lacked a physical infrastructure that helped inhibit the spread of diseases. Wooden floors, heated rooms, glass windows, and adequate food sources would have helped the pigs on both farms survive the winter, but they were neither present nor planned for. Perhaps these details were so mundane that they escaped the notice of Soviet planners, who focused on more dramatic new inputs such as purebred sows and visiting American professionals, or perhaps these were details that a farm run by terror could not attend to, dependent as these elements were on stable workforces and a system of accountability between managers and personnel.

In late March, Heikens was reassigned to a purebred stock

farm closer to Moscow, a change he welcomed, in part because it entailed less responsibility and less contact with untrained workers. Heikens noted when he arrived at his new farm that "there are about half as many hogs here . . . so it is lots easier. I do not have enough to do in fact."[40] The Soviet government urged Heikens to renew his contract, but offered less tempting terms; his salary would be paid out only in rubles, which were nearly worthless outside of the Soviet Union. By October 1931, Heikens was back to working his family's farm in Spencer, Iowa. Guy Bush also chose not to renew his contract. He remained on Kolkhoz 22 until the summer of 1931, and then traveled home through Germany, Denmark, and France, arriving in America in the fall of 1931.

Documents from Central Committee meetings reveal one more way in which there was a gap between what Soviet state officials planned to accomplish on collective farms and the reality of failed and reduced expectations that they were forced to accept. Before Guy Bush and George Heikens arrived in the Soviet Union in 1930, the Party's Central Committee was aware that foreign experts were inefficiently utilized, that their morale was low, and that they almost never renewed their contracts. One report, "On the Improper Usage of Foreign Construction Workers," noted that the first 125 foreign workers the Soviet Construction Trust hired had arrived unprepared to assume leadership positions and that their work skills were underexploited; for instance, an expert in cement spent weeks working as an unskilled line worker. Foreign workers were also not used to the everyday hardships that were commonplace in the early years of the Soviet Union. The report noted that they missed conveniences such as pillowcases, drinking water, and washstands. "Without satisfying food, without newspapers, without regular mail from their home country . . . these workers develop a counterrevolutionary attitude and try hard to return to their home countries." The committee noted that two agricultural experts who had served during 1929 had fared especially poorly, and that these experts had not been managed well by their Soviet supervisors.[41]

The Central Committee had numerous proposals for how to solve these problems in 1930 with the new, much larger group of engineers and experts who would be working in the Soviet Union. These included hosting a ten-day reunion in Moscow, providing translators, and ensuring that foreign guest workers had priority access to scarce luxuries like soap, coffee, and winter clothing. Indeed, some of these measures are evident in the letters of Bush and Heikens, but as with the rest of their experiences, the distance between what was proposed and what was accomplished was significant. Both men were assigned translators, but neither had good luck retaining them. George Heikens had four different interpreters in the first three months he lived at Rodomanovo. The first three found life on the kolkhoz too uncomfortable and resigned their positions. Both Heikens and Bush had access to the best supplies of food, clothing, and other personal supplies in the Soviet Union, as the Central Committee had recommended. However, this allowed for only a slight amelioration in their quality of life. In the dire food situation in the Rostov region during the winter of 1930–31 Guy Bush's preferred status allowed him to purchase only a single sheep carcass, which lasted for about six weeks, though it also ensured that he had plentiful amounts of the same soup and bread his Soviet counterparts consumed in more modest helpings.

In August of 1931, Heikens, Bush, and most of the other

foreign livestock specialists posted across the European Soviet Union convened in Moscow for the ten-day reunion that the Central Committee had proposed to boost morale among foreign workers. Bush noted that at the meeting, the specialists had heard a presentation about hog farms planned for the Second Five Year Plan:

> [It] calls for massive buildings housing 5,000 or more swine each. The one proposed was 1/2 mile long, two stories high and the pigs were to be fed from conveyor belts. . . . the project was so different that one of the Americans in a satirical mood designed a four-story hog house. The roof was to be used as an exercise lot for sows. Each floor was to have a trap door that would automatically drop a hog to the lower floor when it reached a certain weight until eventually it came out as sausage. In the mechanism the exercising sows were furnishing power to operate the sausage mill. Needless to say the latter project was not presented but of course was widely discussed by the four of us in attendance—especially after a little liquid encouragement.[42]

Some of the boost in morale that this meeting might have had for foreign specialists came at the expense of their Soviet hosts and their ambitious plans. Americans might joke privately about the Second Five Year Plan's overautomation and outsized scale for hog farms, but these jokes contained an implicit recognition that these plans seemed impractical and unattainable.

The work of Bush and Heikens had little positive impact on the Soviet swine population. There were not significantly more pigs in 1932 than there had been in 1931 or 1930. But the symbolic effects of hiring outside professionals lingered, and Soviet agricultural planners would rely on outside authority and foreign, largely American standards of efficiency and excellence in agriculture for much of the next three decades. Heikens and Bush deferred to other outside authorities as well: Heikens subscribed to no fewer than six U.S. agricultural publications while he was in the country, including the journal Bush had worked at, *Wallaces Farmer*. This deference to professional authorities was a hallmark of twentieth-century agricultural professionalization in the United States. It was this model that the Soviet Union tried to import in 1930, and to which the country would return after the Second World War.

Heikens and Bush expected to help rapidly construct a modern system of animal agriculture in the Soviet Union during their time there, but the high salaries and privileges given to foreign workers by the state were not enough to outweigh the barriers to progress that existed on new kolkhozes. Although their postings were hundreds of miles apart, the two men encountered similar sticking points on the farms where they worked. And although Bush and Heikens were naïve visitors to the Soviet Union, who witnessed the workings of their respective farms without completely understanding the social or political meaning of collectivization for Soviet farmers or the state, their observations about bad managers, material scarcity, and bureaucratic disorganization on the farms closely echoed accusations made by regional authorities and the Central Committee during 1930 and 1931. Finally, the persistence of a high mortality rate among the animals they cared for eroded morale and ruined the growth and progress of the farm that the Soviet Union was so keen to chart.

Ultimately, Soviet approaches to improving animal agriculture disillusioned both men. In order to expand, farms needed to be managed in a way that could increase animal populations, but livestock required careful management in order to thrive. A prospering Soviet pig, sheep, cow, or horse in 1930 needed an abundant and well-planned diet, access to competent veterinary care, and relative freedom from the harsher elements of the environment. To Bush's and Heikens's growing frustration, animal farms in the Soviet Union were not able to offer any of these necessities to their charges and, as a result, postcollectivization farm animals continued to live preindustrial lifestyles, continually failing to meet their state-planned targets. Farm animal populations stagnated throughout the 1930s.

In some ways, the Soviet Union's experiences creating animal farms were less anomalous than those of the United States, a country that had industrialized its fields and barns in less than a generation. Guy Bush and George Heikens were occasionally shocked by the primitive conditions of the farms on which they worked, but both men were the products of an economic system and an environmental milieu that had made industrializing easy. American successes in raising grain crops, hogs, and cattle were built upon the rare U.S. confluence of a relatively temperate climate, cheap land, and free water. Alfalfa, an ideal animal feed, grew abundantly, and once farm machines such as tractors and combine harvesters helped overcome the chronic labor shortages that had plagued nineteenth-century American agricultural expansion, the result was an explosion in agro-industrial productivity. Many American farmers chose to raise animals as a "value-added" product; they could fatten livestock on cheap grain. Government subsidies helped keep meat and milk prices low for consumers, and by the turn of the twentieth century Americans ate more meat and drank more milk than citizens of almost any other nation. What worked in the United States did not necessarily work everywhere, however, as Americans learned at several points during the ensuing decades when their agricultural professionals made earnest attempts to export the American model of farming to other countries.

The Soviet Union's very different climate did not lend itself to surplus agricultural production. Cold weather limited the productivity of cereal grasses, especially wheat, and specialty crops such as alfalfa required either abundant summer rainfall or reliable irrigation, neither of which was typical of the country. Feed crops available in the Soviet Union in the 1930s included roots such as turnips, potatoes, and rutabagas. Guy Bush repeatedly requested alfalfa, but there was none available for Kolkhoz 22 in the spring of 1931. The Soviet Union also had very different priorities for its labor sector; a primary concern in the Soviet Union was ensuring that collective farms were able to offer employment to every kolkhoz member who wished to work. Although the Soviet Union dreamed of mechanizing all aspects of farm and factory labor, these dreams conflicted with several realities. First, the country had an abundant labor pool. Throughout the Soviet period, a higher proportion of Soviet citizens lived and worked on farms than in almost any other industrialized country. Secondly, agricultural mechanization was difficult to accomplish in one fluid motion. By the end of the 1930s, many kolkhozes had access to tractors, but the infrastructure required to industrialize beyond the use of sophisticated farm equipment often eluded the Soviet state. From the time of collectivization until the end of the Soviet period, there was a disconnect between the Soviet Union's much-celebrated dream of total mechanization and automation, and the reality that, in agricultural settings, replacing human labor with machines was often not the best option. Particularly during the 1930s, Soviet agricultural landscapes contained a surplus of human workers and a dearth of both machines and their necessary support networks, including diesel fuel, skilled mechanics, and spare parts. Later in the decade, Machine Tractor Stations would partially eliminate this lack of technology, but the Soviet state would need to return to the challenge of employing and supporting its rural populace numerous times over the course of the next three decades.[43]

The state-sponsored famine that ravaged the Soviet Union between the winter of 1932 and the summer of 1934 is the most memorable event linked to Stalin's mass-collectivization drive, but it was not the only consequence of collectivization. The testimony of the two American experts and the handful of archival documents presented here emphasize that the legacy of collectivization was not a single tragic event, but the introduction of a new system of control and a new hierarchy of power in rural communities across the Soviet Union. Self-governance was abolished, replaced by a bureaucratic chain of command. Small, privately owned farms with diverse harvests and high levels of self-reliance merged to become state-managed enterprises, often specializing in raising only one or two cash crops. Where it was practical, tractors replaced hand labor in fields and experts such as Bush and Heikens created and supervised new farming protocols that were intended to raise agricultural productivity. All of these interventions were ways in which the Soviet state tried to both industrialize and control its vast, unruly countryside in one swift action.

Bringing its rural areas under control proved to be an almost impossible task for the Soviet state, and collectivization was plagued with problems at every level from the very beginning. Protesters against collectivization killed many of the Soviet Union's already scarce supply of livestock. Most farm managers were incompetent, and this hindered productivity. Kolkhoz workers resented the state's heavy-handed interventions, which further hindered productivity. Crucially, the Soviet state overlooked its primary adversary, which was natural, not cultural. The Soviet state approached Soviet rural areas as socially backward places that needed order and discipline in order to become modern, but agriculture and daily life in the Soviet Union were dominated by austere natural realities far more than by backward social mentalities or revolutionary political spirits. Humans mattered in the Soviet rural landscape, and the state's wanton disregard for their well-being is an important story. However, in terms of the success or failure of state-sponsored interventions on farms in the Soviet Union, nature mattered more.

In the cases of Rodomanovo, Kolkhoz 22, and countless other new collective farms scattered across the Soviet Union in 1930 and 1931, nature obstructed progress by fostering diseases, making crop yields unpredictable, creating harsh conditions in which it was difficult to keep animals alive, isolating villages from rail lines and main roads for months during the winter, and making outdoor work either uncomfortable or impossible for much of that season. Guy Bush and George Heikens were outsiders, but they quickly perceived the extremely challenging task the Soviet state had set for itself in modernizing the countryside. Their letters recount the chronic shortfalls and mistakes the state made in these very early years, and their critiques resonate with the internal criticisms the Central Committee and local officials made about agriculture during the same time period. The gap between what the Soviet government wanted to achieve in the newly collectivized countryside and what it was capable of accomplishing was vast in 1930 and 1931. Over the course of the next thirty years, this gap would eventually close, but it is important to understand just how far the Soviet Union was from realizing its dream of efficient industrialization on its farms in the aftermath of collectivization.

2

Restoring Control

State-sponsored interventions in the 1930s introduced scientific and bureaucratic professionals to Soviet farms, but many of these professionals did not remain in rural posts very long. Some, like the veterinarian on Guy Bush's farm and the first farm manager at Rodomanovo, lost their jobs because of poor performance. Professionals could be exiled, imprisoned, or even executed for gross incompetence as well as naïve mistakes. Other experts accepted only temporary work contracts and never planned to remain in the countryside. This was the case for George Heikens and Guy Bush, but also for thousands of Soviet urban professionals who formed a short-term village vanguard between 1930 and 1932, a group known as the Twenty-five-thousanders, named for the target number of volunteers recruited into the program.[1]

Whether visiting foreigners or native professionals, outside scientists, engineers, and bureaucrats often inspired resentment on newly formed kolkhozes and their presence failed to inspire monumental change. At the start of collectivization, the state had approached rural underdevelopment as a problem that could be solved with money and industrial streamlining, but in spite of high capital investments and an influx of efficiency experts, the countryside did not become dramatically more productive or less wasteful during the 1930s. Perhaps because of this initial false start toward agricultural industrialization, the state approached modernization very differently after the Second World War. Although the Ministry of Agriculture still relied on outside professionals to steer the countryside toward modern practices in the postwar era, state agencies also focused on institution building and emphasized the areas of agricultural production (for example, transportation and land surveys) where it was possible to achieve almost complete oversight over the countryside. These targeted techniques were more successful, and they became the hallmark of Soviet agriculture for the next twenty years. The start of this improved, more selective postwar approach to rural governance is the focus of this chapter.

The Second World War devastated the Soviet Union. Although victorious, it endured more losses than any other nation that fought in the war. One in five adult men never returned at war's end, and another 10 percent of the population was permanently physically disabled. As a wartime defensive measure the Red Army had burned and destroyed crops, machines, and buildings as it retreated across western Russia and Ukraine. Postwar reconstruction in these areas was slow, especially in rural districts. Unlike Western Europe, the Soviet Union received no financial assistance from the United States or other countries to finance rebuilding. Stripping parts of Eastern Europe and the occupied Soviet Zone of Germany of its industrial base and materials did little to make up for the enormous devastation war caused in rural areas.[2]

After the Second World War, the Ministry of Agriculture

needed to restore its authority across a countryside that had been largely neglected or laid waste during the war.[3] This task proved to be more challenging in some areas than in others. Rebuilding showcase farms such as those where Heikens and Bush had worked was not complicated because the government was already heavily invested in these facilities, and as a result the ministry prioritized reasserting control over such valuable assets. The real challenges came from the much more common smaller and more marginal collective farms scattered across the Soviet Union, especially those in southern Russia and Ukraine that had been occupied by the German army during the war. These farms had been collectivized in the 1930s, but they had been neglected and left to their own devices for the duration of the war.

Like most projects devised by the Ministry of Agriculture, rebuilding marginal farms after the war was more easily planned than executed. The bureaucratic infrastructure the state had created in the countryside during the 1930s had vanished during the war. Purges and dekulakization had removed some of the most competent managers from the countryside in the late 1930s before the war started, and many other farm administrators had left the countryside to serve in the armed forces when war broke out. During the war many collective farmers reverted to subsistence agriculture that required no outside management. This shift in cultivation strategy proved to be good for individual self-preservation, but bad for collective national food production. The Soviet countryside of the late 1940s was a more independent and less organized place than it had been during the 1930s, and the Ministry of Agriculture was deeply concerned because rates of agricultural production were much lower than they had been before the war.

Re-creating the state of rural order and control the central government had achieved in the Soviet countryside during the 1930s was a significant challenge for the postwar Ministry of Agriculture. This was especially true in Ukraine, where many citizens retained a strong sense of Ukrainian nationalism and resented the overbearing, federated Soviet state. Most of Ukraine had been occupied by Germany during the war, and this inspired deep (and not entirely unfounded) suspicions on the part of the Soviet government about the loyalty of Ukrainian citizens. Ukrainian farmers were reluctant to return to centrally organized collective forms of agriculture in the postwar era, in part because at least some of them were reluctant to rejoin the Soviet Union at all.

Nevertheless, the state was able through various mechanisms to reassert control over the smaller, less profitable farms that comprised much of its agricultural portfolio. These were not the high-profile livestock farms that had hired Heikens and Bush; they did not specialize in raising animals for meat or milk. Rather, they raised grains, vegetables, and potatoes, the foundational ingredients of the Soviet diet. The Ministry of Agriculture used claims about its duty to control and manage agricultural diseases, pests, and invasive species to reclaim its power over such smaller kolkhozes. By alleging that these farms were overrun with a variety of diseases and pests, the Soviet state increased its power in parts of Ukraine and reaffirmed and strengthened its role as protector and overseer of the country's farmland.

The role of kolkhozes began to change as well. Previously they had been the principal economic unit of socialist agriculture, but in the postwar period they became less important than the larger and more industrialized sovkhozes. In the postwar era, the state no longer relied on kolkhozes to produce the bulk of wheat and other staple grains; instead they were earmarked to produce agricultural surpluses that could be used to make processed foods and animal feeds, while sovkhozes gradually assumed the task of feeding the nation and producing grain for export. In the late 1940s, this transition was just beginning to gather momentum.[4] These twin tactics of the Soviet state—regaining and refining bureaucratic control by creating and expanding scientific institutions on one hand, and creating parallel and unequal farming structures consisting of smaller kolkhozes and larger sovkhozes on the other—were the central tenets of Soviet agricultural policy during the immediate postwar period.

The Second World War damaged the Soviet countryside, but rural residents also remembered the war as a time when they were left alone by the government after the many abuses of the 1930s. In the urgency of wartime, the state stopped intensively managing the countryside and focused instead on winning the war against Germany. State neglect was not unwelcome, although it often resulted in extreme adversity. Regions that fell under the control of the German army did not always feel as if they were being oppressed by an enemy; in postwar interviews respondents who had worked on occupied farms felt that the Germans ran kolkhozes in nearly the same way the Soviets had done, except that villagers were allowed to practice religion in public.[5] Collective farmers who lived in villages that remained unoccupied relied on private-sector markets, illegal trade, and individual garden plots to furnish the food they needed to survive. The war years were a period of extreme privation for many Soviet citizens, and civilians starved to death in a number of cities and regions that were occupied or blockaded by the German Army.

One wartime food policy that had far-reaching implications for the postwar period was the state's allocation of private allotment gardens to workers. In 1942 the state began assigning plots to state employees, including kolkhozniks. It also turned a blind eye to unsanctioned private cultivation; in wartime the government could not really afford to discourage any form of food production, however individualistic. This reversed a decade of earlier policies that had discouraged cultivation on private plots. From 1929 until the German invasion in 1941 the state had systematically worked to reduce access to privately produced food by imposing taxes on land and private production, by claiming ownership of privately held land, and by razing illegal private garden plots. Over the course of the 1930s the amount of food Soviet citizens produced outside of collectives dropped significantly. This takeover of private plots—strip farms as well as backyard gardens—had motivated the collective farm revolts of March 1930. The right to cultivate an allotment was one of the few concessions farmers won back from the central government in the aftermath of collectivization.[6]

Throughout the 1930s, kolkhoz workers both resented and ignored the restrictions that limited private production, in spite of high tax penalties, public shaming efforts, and a lack of appropriate tools (which had been confiscated in collectivization drives). They continued to work in allotment gardens and to privately sell their produce. In this context, gardening was an act of resistance and a quiet vote of no confidence in the state's claim that it could reliably and justly redistribute food products. Collective farmers had lived through famines in 1921 and 1932–34, and they saw no reason to trust the Soviet state as a provisioning authority. Throughout the 1930s collective farmers with gardens provided basic foodstuffs such as melons, turnips, cabbages, beets, and potatoes, as well as scarcer luxury food items such as chickens, butter, sour cream, and eggs. Selling these products at informal markets was one of the few ways in which peasants could acquire rubles rather than payments in kind in exchange for their labor.

Prewar gardens were taxed heavily and kolkhoz members had little time to work in them. One former kolkhoz worker who sought refuge in occupied Germany after the Second World War II remembered that a brigade system of labor was introduced on his kolkhoz as a punishment when families spent too much time on individual gardening tasks and not enough time on their assigned work in the collective fields. The war reversed state policies on private production and by November of 1942 Soviet citizens were encouraged to grow as much food as possible on allotments and in public spaces in order to help in the war effort. While allotment gardens did not solve the chronic civilian food shortages associated with the war, they did make up for a significant portion of the loss, especially in the occupied lands in Ukraine and southern Russia.[7]

The state actively encouraged private cultivation as a wartime emergency provision, but these policies were supposed to change at the end of the war. The Ministry of Agriculture planned to transition back to controlling private production, but kolkhoz workers recognized their marginal position in the immediate postwar period and refused to stop growing food to both eat and sell. This instinct of self-preservation was complemented by the euphoria that victory had brought to the nation as a whole; many people anticipated that Stalin would now allow wider freedoms and a better economic life.[8] Contrary to popular expectation, postwar state policy continued many of the most repressive and paranoid aspects of the Stalinist dictatorship; survivors of the German occupation were suspected of antistate activities, and in the immediate wake of the Holocaust, Stalin instigated an anti-Semitic campaign that targeted Jewish professionals. A postwar famine in Ukraine diminished resistance to state control in the region and served as an early warning that the Soviet state was still willing to withhold food as a way of coercing the populace into doing what it wanted. In spite of these postwar signals, however, in rural areas the Soviet government was not able to reassert control over many aspects of daily life immediately after the war. This was especially true in small towns and on kolkhozes.

Previously, in the early 1930s, new kolkhoz bureaucracies had consisted of one or two repurposed village buildings (often churches), a crew of surly and inexperienced workers, revolving-door managers, and a sporadic, biased system of performance review. In 1930 and 1931 almost every worker on a kolkhoz worked twelve or thirteen twelve-hour days out of every fortnight. In this early system frenetic activity and an influx of outside experts were supposed to substitute for the efficiency and organization of an experienced and well-established farming operation. After the Second World War the Ministry of Agriculture, more experienced and also more realistic about its abilities to organize and manage rural areas, used a different set of tactics to recollectivize the parts of the Soviet Union that had been occupied during the war. Kolkhoz chairmen had less power in the postwar period and the Ministry of Agriculture relied more on stable scientific and technical bureaucracies that hovered at the margins of rural life. Machine Tractor Stations, which coordinated the renting of tractors and other heavy machinery to kolkhozes, were one such authority.[9] Another important institution was the scientifically focused Quarantine Stations, whose role was to inspect plants and seeds for pests, and to prevent communicable plant and animal diseases from entering or spreading within the Soviet Union. The Ministry of Agriculture found it easier to stock these larger institutions with a stable population of trained professionals, and depended less on reliable self-governance from each individual kolkhoz, a lesson it had learned in the 1930s with its struggles to appoint and retain a competent class of managers on its collective farms.

A few sovkhozes had been established already in the 1920s, but their numbers expanded rapidly after the war. Sovkhozes differed from kolkhozes in several important ways. In the first place, the state owned the land, barns, houses, and other buildings associated with the farm, whereas on kolkhozes, the local community and not the state shared ownership of these kinds of property. This meant that sovkhozes functioned more like urban factories than rural farmsteads. Workers slept in dormitories, received cash wages, and frequently changed jobs or transferred away from their sovkhoz, work patterns that were uncommon on kolkhozes. Sovkhozes also tended to specialize in just one or a few products, and benefited from economies of scale. Because they resembled industrial units and because they were larger and more efficient than kolkhozes, state policies favored sovkhozes. They were first in line to receive tractors, combines, and other machines that might improve productivity, and they were the first farms to be given improved new seeds and purebred and other high-producing animal stocks. In general, the sovkhoz was perceived by the state as a more reliable institution because the government owned it. By getting first preference on new equipment and new supplies, state farms were set up to succeed in a way that collective farms were not. Furthermore, cash wages gave sovkhoz workers more incentive to perform, whereas kolkhoz members received workday credits, the perennially unpopular *trudodni*. Workday credits were detested because they rarely translated into money that could be used off-farm; instead they were typically traded in for a certain amount of grain. In essence they were a form of state-sponsored scrip that reinforced the dependency between collective farm workers and their employees, binding kolkhoz workers to the farms as the sole supplier of food and other essentials and limiting purchasing options.

For the first three years after the war, in parts of formerly occupied Ukraine and Southern Russia, kolkhoz workers refused to switch from individual subsistence agriculture back to the market agriculture that better served the interests of the state. The majority of these recalcitrant farmers grew potatoes to eat and sell, and they tended their potato crops in lieu of working in state grain fields. Although farmers were willing to sell surplus produce to state-owned subsidiaries for cash (typically selling potatoes to vodka distilleries), they did not want to put much time or labor into collective grain fields. In this early postwar period kolkhoz workers retained control over the kind of crops they raised, the amount of time they worked, and the style in which they farmed. In parts of Ukraine during the war, farmers returned to—and stuck with—strip farming, a technique the Soviet government had condemned as inefficient and backward, and had all but abolished during collectivization.[10]

The ongoing problem of weed infestations on kolkhozes in the postwar era was one way in which the reluctance of collective farmers to return to their previous duties was made visible. Farmers abandoned wheat and rye fields throughout Ukraine during the German occupation, and when these fields were replanted after the war, they were overrun with thistles, which limited their productive capacity. This was a serious problem, but one that could have been easily solved by a single season of weeding. Yet in spite of advertising campaigns that called for a critical mass of workers to weed fields, the removal of thistles—a job that could be performed by children—was unsuccessful until at least 1949.[11] The postwar countryside had no lack of people, but there was not a population of reliable workers who were willing to devote sustained energy to a simple but time-consuming project such as weeding a field that was certain to be planted for grain. The Soviet state had few tempting incentives with which to persuade peasants to abandon their gardens and return to collective fields. Inducements had never been the Ministry of Agriculture's strong suit. In the postwar era a typical kolkhoz member worked harder, earned less money, ate less, and was more susceptible to the natural perils of farming than he had been as a freeholding peasant during the New Economic Policy of the 1920s. Not performing an extra task like weeding was one of the few ways in which such a worker could assert his nominal autonomy.

The war also dissolved a number of effective, if heavy-handed surveillance techniques that the state had adopted during the 1930s. In these policing methods, loyal outside bureaucrats inspected farm work periodically, especially during harvest time or if a farm failed to produce as much grain or other commodity as expected. This was one of the key tasks performed by the resented Twenty-five-thousanders who descended on Soviet villages for a few months in an attempt to improve kolkhoz efficiency.[12] Strong-arm tactics were not limited to volunteer outsiders; local security teams guarded collective fields against theft from other farmers, and kolkhoz members who hoarded grain or stole collective resources were punished harshly in public. In wartime this high level of vigilance disappeared and it did not return at war's end. Thus there were few incentives for peasants to engage in collective work, and there were no threats of punishment if they failed to do so.

Although the continued independent choices of cultivators in Ukraine and other parts of the Soviet Union posed a

significant challenge to the Soviet state, the postwar agricultural landscape was not one of abject statelessness. The Ministry of Procurements purchased or requisitioned 17.5 million
tons of grain in 1946 and 27.9 million tons in 1947. Indeed, this
level of state intervention helped to create a serious shortage of
food in many parts of Ukraine and southern Russia. Overzeal-
ous state grain collection policies coupled with a devastating
drought across Ukraine and southern and eastern Russia created a serious famine in the region. The Ministry of Agriculture
and other central authorities covered up evidence of this famine for decades, in part because it was further evidence of the
state's inability to perfectly regulate agricultural landscapes.
The state also exacerbated food shortages across drought-
stricken regions by failing to provide palliative or emergency
foods to rural residents, reducing existing rations to cities and
towns, and increasing its stockpiles of grain throughout the
most severe period of the famine, in 1947. It also blocked outsiders from making accurate observations in the region.[13]

Although a few contemporary documents describe the
postwar famine in Ukraine, the eastern Urals, and Kazakhstan,
they are rare. Instead, this famine is most evident in what is missing from the historical record. For example it is now known that
between one and two million—predominantly rural—people
died—a figure that is based not on deaths formally registered
in those years, but on the estimated population that was missing from census data in famine-affected parts of the country
after 1934. The lack of information about affected regions was
not just oversight, but a planned policy decision for the Ministry of Agriculture, which cut off foreign and domestic access
to areas of the country affected by food shortages in 1946. For
example, the monthly Narrative Report of the U.S. Foreign
Agricultural Service in the USSR cuts off abruptly in the midst

of the 1946 harvest. In 1946 and 1947 combined there are only three reports on agricultural conditions in any part of the country, which differs significantly from the pattern of regular reporting established both before and after the famine.[14]

American officials who witnessed the beginning stages of the famine also declined to sound an alarm. In 1946, one agent, Joseph Bulik, wrote the majority of reports for the U.S. State Department about crop conditions across the Soviet Union. Because of widespread drought and general postwar hardship, many of Bulik's colleagues in other European countries used their 1946 crop reports to build the case for American food assistance to these regions, describing hunger and human suffering in detail. In the immediate postwar period food was scarce and the United States was on the lookout for signs of famine and malnutrition across much of Western Europe. In contrast, Bulik alluded just once to the impending food crisis in Ukraine, noting that there was "very little margin indeed for the peasant until the 1946 harvest could be taken in. . . . This . . . has created the kind of situation not conducive to the current happiness of a hard-headed peasant as concerns his local and central government." In the same report, Bulik mentioned that one woman working in the fields in Ukraine spontaneously cried out to him as he stopped to take her photograph, "we are dying of hunger!" Bulik recounted her words without offering any commentary on their significance or context.[15]

Bulik did not bother to document the likelihood of a Ukrainian famine in part because by 1946, American economic assistance was already earmarked for other countries. President Truman had withdrawn most material assistance to the Soviet Union in the summer of 1945, and there was nothing to indicate that the country would regain this aid. Even before the war ended, Averill Harriman, the U.S. ambassador to the Soviet Union during the war, wrote that he had "come to the conclusion that we should be guided as a matter of principle by the policy of taking care of our Western allies and other areas under our responsibility first, allocating to Russia what is left. We should . . . reestablish a reasonable life for the people of these countries [Greece and Italy] who have the same general outlook as we have on life and the development of the world. The Soviets . . . have an entirely different objective."[16] It was in America's best interest to ignore the looming famine in 1946, regardless of how apparent it might have been that people were on the verge of starving. Had U.S. officials like Bulik mentioned a critical food shortage, it would have contradicted official Soviet reports that the country had collected a successful harvest in 1946. It also might have ethically obligated the United States to act to ameliorate food shortages, an expensive and diplomatically troublesome action. It was both easier and more politically expedient for Joseph Bulik and other American observers to ignore early signs of famine and instead concur with the reports of politicians in Moscow that there was plenty of food, and that the state had the situation under control.

The United States and the USSR competed over national agricultural capacity after the war, a fact that influenced both official Soviet reports and U.S. estimates of the Soviet Union's crops. The United States and Canada harvested bumper grain crops between 1945 and 1947, but both countries suspected the Soviet Union had underreported its anticipated wheat crops in order to keep world prices high until its harvest, which was gathered later in the season, could flood the market.[17] The Soviet Union had done this before, during the 1920s, but in this instance American suspicions were unfounded; in both 1946 and 1947 the Soviet wheat harvest was actually much lower than expected. There was no secret bumper crop.

In 1932 and 1933 between five and seven million Soviet people had perished in a famine that was—at least in part—created through stringent state grain requisitions. American diplomats and agricultural attachés working in the Soviet Union after the Second World War worried that new grain demands might again create such conditions. Grain requisitions still sought to redistribute food in order to guarantee city residents access to bread and other staples, but after the war they were neither as demanding nor as punitive as those that had been in place in the early 1930s. Certainly, requisitions still removed grain from the villages that could have been used as local reserves, mitigating at least some of the effects of the two poor harvest years.[18] The 1946–47 famine might have been averted if grain had not been requisitioned *and* the country had been offered food donations such as those available through the Marshall Plan to Western Europe, but Soviet-American relations in the immediate postwar period were at a low point, and outside assistance from the U.S. or Western Europe was never an option. The severe drought in the USSR during the spring and summer of 1946 created conditions conducive to famine, and Cold War animosities and Soviet requisitions made it much more lethal, especially for rural people, and above all for rural Ukrainians.

Grain was not the only staple that Soviet citizens relied upon in the postwar era. Potatoes were an alternate crop that could not be easily requisitioned by the state. This was one reason they were a postwar crop of choice. Historically, potatoes have been a subsistence food because they are not an ideal crop for centralized collecting or long-distance hauling, but the postwar period was the first time in Soviet history that potatoes became a subsistence food. Compared to other parts of Europe, potatoes had arrived relatively recently in Russia; they had been introduced in the late seventeenth century, and had never been widely cultivated. For two centuries, Russians grew the potato as a garden vegetable and occasional animal feed, but not as a staple food, and it did not make a major impact on national agriculture until the twentieth century. In general Russian peasants adopted potatoes as a crop slowly, proceeding with caution in a marginal northern environment.[19]

Potatoes thrive on land where grain does not grow well—in poor soils and in fields choked with weeds. They are also the crop of choice in areas where farmers need to sell cereal crops for cash, but need a staple food to keep to consume during the winter. Historically in Russia and Ukraine, wheat, oats, rye, and millet were all cold-hardy grains that were easy to store and transport. Farmers raised these crops both to sell and to retain as winter food sources. Collectivization altered this model; kolkhoz workers were ordered to turn all harvested grain over to the state, and then they were promised that the state would give some of this grain back as payment in kind. Until the start of the Second World War peasants working on collective farms in the Soviet Union received a large portion of their salary in grain. In 1945, when the United States unexpectedly revoked the postwar food aid it had pledged to the USSR as part of the Lend-Lease program, the Soviet Union had to lean harder on its internal resources, and drastically reduced grain-based payments in kind to Soviet farmers. Kolkhoz workers had already learned they could not rely on state allotments of wheat for winter subsistence, so they turned to potatoes as a survival crop.[20]

The historian Zhores Medvedev claims that potatoes became "a second bread" to Russia only during the Soviet period. In his words, "the only agricultural success which can be attributed to the Soviet system is the increased production of potatoes."[21] Medvedev overstates the failure of the Soviet system in the realm of agriculture, but he accurately notes that the policies of the Soviet state led directly to the adoption of potatoes as a staple crop. The instabilities caused by the Soviet regime and war as well as an overly zealous system of requisitioning grain inspired rural citizens to rely on potatoes as a new staple food.

Some of the aspects that originally prevented the widespread adoption of potatoes in the Soviet countryside were actually instrumental to its success during and after the war. For example, potatoes were not a machine-cultivated crop. Well into the 1950s, potatoes were almost exclusively planted, dug, and harvested by hand. The drudgery of manual cultivation weighed against the state incorporating potatoes into its plans for kolkhoz production. Agricultural planners were biased toward crops that could be mechanized, but during the Second World War the labor liability inherent in potatoes turned into a benefit. When animal draft power and farm machines disappeared or broke down, the burden of planting, guarding, and sowing wheat and other grains substantially increased, whereas potatoes offered higher returns for labor investment when the only power source available was manual labor. Potatoes were also a seasonally flexible crop. Unlike most grains, which have narrow windows of time during which they can be sown and harvested, potatoes could be planted at any point after the threat of frost had passed, roughly from May to July, and harvested either before or after the frosts returned in midautumn. Like turnips, potatoes made a good feed crop for animals, but they also yielded more calories for the amount of work put into the crop than turnips did. After spending six months in a root cellar, potatoes were also considered to be more palatable for human consumption than turnips. Finally, while hungry farmers were probably not considering the vitamin enrichment levels of their staple crops, in hard times potatoes were one of the best nutritional choices among staple foods, especially when compared to the other options available to rural workers in the Soviet Union.

In the matter of storage, the aftermath of war clearly favored potatoes above grain. Potatoes were easier to store than wheat. Late-harvested potatoes, if properly cellared, could keep until early summer, providing a food to tide peasants over during the hard fieldwork of late spring, typically the hungriest part of the year in the agrarian cycle. Further, potatoes were best stored in root cellars rather than in raised wooden granaries, like wheat. German destruction during wartime occupation and Soviet scorched-earth tactics had left few buildings standing in parts of Ukraine and Russia. This included granaries, barns, and silos, but potato storage pits were often spared. Many people who lived through the German occupation in rural Ukraine inhabited earthen dugouts, *zemlianki,* that bore more than a passing resemblance to potato pits. In time of war, these buildings were functional and durable in spite of being cold, dark, and damp. They could not be burned down, and they were both difficult to destroy and easy to rebuild. The damp and gloomy conditions of dugout houses, which were a source of discomfort and ill health for Soviet citizens for years after the war ended, had the lone virtue of keeping potatoes fresh for a long time. The general pattern of material destruction the war caused did not harm potatoes the way it harmed the security of wheat supplies, animal shelters, or human dwellings.

Medvedev describes the rise of the potato during the war. His observations are worth quoting at length, both for the details he provides, as well as for what he omits from this description:

. . . the potato harvest reached a record level of
95 million tons [in 1948], but this in itself was an
expression of peasant resistance. The state had no
proper system for collecting potatoes. Moreover,
the potato harvest is the last in the agricultural season to be collected. By the time it is brought in (October–November) only two or three weeks remain
before the winter frosts. Unlike grain, potatoes can-
not be taken from the villages during the winter,
since frost kills the tubers. Less than 10 percent
of the potato harvest normally reaches the towns
through the state procurement system (7.2 million
tons in 1948) . . . the rest remains in the villages. . . .
Potatoes became the most important staple food,
and could be used instead of feed grain for livestock, particularly for pigs.[22]

The reason that the Soviet state had no "proper system"
for collecting potatoes was that potatoes are heavy and it made
little sense to transport them across great distances. Thus, by
default, they remained a local crop, and this, in turn was one
reason they were adopted as an animal feed in the years before
the war. George Heikens and Guy Bush both complained bit-
terly in 1931 that pigs on their farms had only potatoes to eat,
rather than more nutritious crops like alfalfa, corn, or soy. They
failed to see that potatoes were useful on small Soviet farms
because they could perform double duty as either human food
or pig feed, depending on the abundance or scarcity of other
food sources. Heikens and Bush were correct that potatoes are
not an ideal food for pigs; the animals could gain more weight
and were more robust when they consumed a higher-protein,
lower-starch diet. However, while optimal porcine nutrition

was a goal of the Soviet state in 1931, there was considerable distance between the ambitions of the state and the reality on the ground. Potatoes kept both pigs and people alive in the postwar Soviet countryside, irrespective of the fact they were not the best possible source of nourishment.

Potatoes became an excellent economic investment for farmers in Ukraine immediately after the war, in part because scarcity increased food prices dramatically. The price of both bread and potatoes had already skyrocketed during the war, and these staples became even more expensive during the leanest postwar years. By July 1943 the price of both foods had risen to over twenty-five times their price in July 1942, and bread prices did not decrease significantly until 1948. Bread, as an expensive commodity made from grain, baked in cities, and paid for with cash, was scarce in rural Ukraine, but potatoes could bring high returns from markets, or they could be eaten if the food supply for the winter and spring was particularly scarce. In some years, farmers could sell excess potatoes at fixed low prices to state-run vodka distilleries, but the high informal market price of potatoes in the postwar years meant that most small growers would get the best price by ignoring the state and selling potatoes as fresh "truck" produce.[23]

The greatest threat the postwar period posed to Soviet agriculture was to the productive capacity of smallholders. The war had destroyed much of the countryside, and in response to food insecurity collective farm workers grew potatoes and refused to work in collective fields. Human reticence, lack of labor power, drought, disorganization, a dearth of machinery and animals, weed infestations, and inadequate distribution networks all stymied postwar agricultural productivity. The Ministry of Agriculture ignored these issues, possibly because they were essentially insurmountable problems for a government agency that had little authority left in the villages. Instead, the ministry focused on more manageable, but arguably much less pressing problems. One such problem was the task of inspecting food for biological contamination. To do this, the state created the Quarantine Stations.

These were not the first institutions of their kind to exist: plant Quarantine Stations had a historical antecedent in the general Quarantine Stations created in Russia during the first half of the nineteenth century in response to cholera outbreaks. These earlier stations were designed as checkpoints where medical officers inspected people and the goods they carried for signs of choleric infection. Until the later nineteenth century, doctors did not completely understand how the disease was transmitted; they simply suspected it was contagious. When station workers found contaminated materials they gassed them with chlorine, and detained infected people until signs of illness passed. In spite of these measures, Russia's imperial government failed to contain or slow the spread of the disease; as one historian of Russian medicine has noted, "whereas quarantine instructions on paper took advantage of the latest developments in scientific medicine, their implementation did not."[24]

This deficit between plan and execution remained true in the second half of the nineteenth century, when the mechanisms of disease transmission through contaminated food and water were better understood. Late imperial quarantine models tried to follow the advice of the German microbiologist Robert Koch, who advocated using medical professionals rather than soldiers to monitor and control Quarantine Stations. Koch's ideas assumed a strong state bureaucratic system, however, something that Russia's tsars endeavored to build during the nineteenth century, but ultimately failed to achieve. This weak state bureaucracy was one major reason for the

Russian empire's failed attempt to contain cholera outbreaks throughout the century.[25]

Cholera killed hundreds of thousands of Russian citizens for decades after the disease had been controlled in Western Europe. It is estimated to have been responsible for 1.5 million deaths in Russia during the nineteenth century—800,000 in a single eighteen-month period in 1893 and 1894—and the disease killed another half-million Russians during the first quarter of the twentieth century.[26]

Reestablishing control over people and goods in the postwar period was at the heart of the postwar Quarantine Station mission. The stations screened agricultural products for infestations and diseases that might endanger the value and productivity of Soviet crops. In the words of the Soviet propaganda office, the rather militant-sounding job of the Quarantine Station system was to "expose and destroy" unwanted pests. The Ministry of Agriculture had created the station system just before the Second World War, but its powers and influence greatly expanded after the war. The concept of agricultural quarantine was not unique to the Soviet Union; the United States had operated a Plant Quarantine and Control Administration for decades, and Germany and Great Britain ran similar facilities in the early years of the twentieth century. The Soviet Union's creation of the Quarantine Station system was part of a larger shift toward the professionalization of agricultural science, and the industrialization and orderly management of the Soviet countryside. In Ukraine, Quarantine Stations were staffed (or in some cases, restaffed) beginning in 1946, although the vast majority of their work was done after 1948.[27]

As a professional scientific institution, the Quarantine Station was basic, with a skeleton staff that had only a few years of university-level training, and primitive facilities. Quarantine

Stations remained a separate bureaucracy from the kolkhozes, under the direct control of the Ministry of Agriculture instead of the indirect control of the local kolkhoz authorities. Quarantine Stations were not concerned with the management of kolkhoz workers. Instead they kept tabs on agricultural products emerging from the kolkhozes. By claiming the authority to inspect both "official" kolkhoz products that had been raised and collected according to the all-important Five Year Plan of the government as well as the unofficial products that farmers grew on the side in order to supplement their meager income and uncertain grain allotments, the state gained considerable influence over these unofficial crops. Influence over this part of the informal agricultural economy had been absent in the years before the Second World War. Because private production was vital for farmers in the postwar Soviet Union, this new mode of surveillance was crucial.

One difference between the Soviet Quarantine Stations and their equivalents in other countries was that the Soviet stations screened and quarantined not just imported goods but domestic food products as well; in fact, in the immediate postwar period, domestic products were the focus of the stations' work. A 1950 directive for one station stated that it was to exert "control over imported and exported material under quarantine, both that within the bounds of the oblast, and outside of its bounds."[28]

The Quarantine Stations had been conceived originally as a practical measure to keep unwanted diseases and pests from spreading, but in the postwar era they also provided the Soviet state with a much-desired way to keep watch over rural workers and the fruits of their labor. One of the four directives of the quarantine system was to organize the "work of *internal* (as opposed to international) quarantine sites" that focused

the brunt of their energy on individual production, such as the
work being done by the potato growers.[29] If workers at a Quarantine Station believed a field or a crop posed a particular risk
to the larger agricultural area, the station had an obligation
to isolate it in order to ensure that its potential infection did
not spread. Although Quarantine Station workers did not generally confiscate potentially contaminated crops, the potato
fields of Ukraine were frequently subject to this kind of intervention between 1947 and 1953. Potato fields claimed by individual farmers and not held under collective ownership were
especially vulnerable to inspection and seizure. In the immediate postwar period, Quarantine Station staff had good reasons for believing Ukrainian potato fields might be vulnerable
to diseases and infestations. Many field crops suffered from
soil diseases, as well as the depredations of grasshoppers and
mice populations that had exploded during the war years.
However, their methods of identifying, isolating, and treating
these fields indicate that their work was fueled by the ulterior
motives of discouraging the growing of potatoes specifically,
and self-production of food in general.

Quarantine Station workers were initially on the hunt for
signs of potato blight, the same fungal infection that contributed to the death of millions during the Irish potato famine of
the 1840s. Soviet officials had good cause to fear the blight in
the volatile period of food scarcity after the war. In Ukraine
a disease that could destroy over half the annual potato crop
could have resulted in famine for tens of millions. Furthermore, potato blight was present in Ukraine, at least to some
degree. In the late 1940s, officials gathered tubers and plants
that exhibited characteristics of the disease and sent them to
the main pathology laboratory at the Plant Research Station
in Leningrad, where they were photographed and catalogued.

The spread of blight was dependent on weather conditions: a cool, rainy spring would encourage the spread of the disease, whereas warmer, drier weather could stop or prevent its spread. Nineteen forty-six and 1947 were drought years, which kept potato blight in check. In spite of the cooperative weather, however, the Quarantine Stations devoted much of their time to searching out blighted potatoes, and to performing experimental and unproductive treatments on affected fields.[30]

In the 1940s, potato blight was treated with chloropicrin, a chemical used as a tear gas in both world wars.[31] Unlike its American and Western European counterparts, the Soviet Union did not possess a surplus stockpile of chloropicrin after the war, primarily because the gas had not been a major weapon used by the Red Army. Although the Soviet Union removed chloropicrin stores from a number of German chemical factories at the end of the war in 1945, moving bulky tanks of chemicals around the USSR was a slow and laborious process and the distribution of chloropicrin was uneven. In 1947 the Soviet Union was neither producing nor importing enough poison to effectively prevent or fight an outbreak of blight. In spite of the fact that chloropicrin was not a practical solution, what small stocks the state did possess were transferred to the Ministry of Agriculture and distributed to Quarantine Stations for experimental use in fighting the blight. Gassing fields became a demonstration project that showed off the anticipated future potential of the Quarantine Stations, much as the pig farms to which George Heikens and Guy Bush were assigned were meant to showcase what the future, rather than the present, of Soviet agriculture would look like.

Although the specter of potato blight was real in the postwar era, the Soviet government also had ideological incentives for investing in the services of a surveillance authority. In its original invasion of the Soviet Union, the German army had occupied most of Ukraine in a matter of weeks, a surprisingly swift operation that had been a major blow to Allied morale, especially to the Red Army. While this was mostly due to the fact that the Red Army was initially slow to mobilize against the threat of German invasion, rumors persisted, alleging that many Ukrainian peasants had betrayed their country and welcomed the German occupation, seeing the invasion as an opportunity to renew a nationalist movement. After the war, the borders of Ukraine expanded when the USSR claimed formerly Polish territories. In the aftermath of reoccupation, with a newly expanded territory and lingering fears of a Ukrainian fifth column, the Quarantine Stations' dual focus on border control and smallholder surveillance fulfilled the central government's desire to reestablish definitive control over the Ukrainian countryside. While there was no indication that the potato blight Soviet scientists identified in Ukrainian fields in the late 1940s was anything other than endemic, the Quarantine Stations' official position was that it had been imported into Ukraine during the German occupation (presumably by German forces).[32] Soviet concerns about the potential permeability of political territory carried over into the Ministry of Agriculture's beliefs about national ecology: unwanted pests and diseases were foreign, Western invaders, and they were labeled as such in press releases and official reports.

It was not unreasonable for the Soviet government to worry that its potato fields might become the object of a biological attack. For the Soviet Union the most dreaded biological agent in the immediate postwar era was not blight or virus diseases, but a North American insect—the Colorado potato beetle. Although there is no direct evidence that the Germans used beetles to attack Soviet agriculture, both Axis and Allied powers built robust wartime biological warfare programs aimed at countering insects that could threaten agricultural productivity. In 1942 Germany established a Potato Beetle Defense Service, responding to rumors that Great Britain already sponsored a secret potato beetle-breeding facility. Initially focused on developing a pesticide that would kill the beetle, the Potato Beetle Defense Service soon became a beetle-breeding facility in its own right. By the summer of 1944, Germany had between twenty and forty million potato beetles on hand. The plan was to fly the beetles to western regions of Britain and release them above potato fields. The Germans hoped the beetles would land safely and demolish the major food staple of the British diet. The plot was never carried out because in local trials, scientists could not recover potato beetles dropped from airplanes. While this may have meant that airdrops widely dispersed the beetles, it was also possible that the flightless insects perished when they hit the ground. Whether Germany decided to cautiously husband its beetles or lost enthusiasm for creative offensive sallies by the summer of 1944 is not clear from the historical record, but mass beetle drops in enemy territory never took place.[33]

As the name suggests, the Colorado potato beetle is native to the western Rocky Mountains in the United States. Its scientific name is *Leptinotarsa decemlineata,* on account of the ten bright yellow stripes that appear on its forewings. Until American farm settlements arrived in the West, the potato beetle's primary food was a relative of the potato, *Solanum rostratum,* a nondescript member of the nightshade family. Its small leaves and poisonous spines meant that the potato beetle was its only major predator, but its scarcity and the great distance that separated individual plants kept beetle populations in check. Settler cultivation of *Solanum tuberosum,* the potato, began in the 1850s. It took *decemlineata* thirty years to adapt to *rostratum*'s domesticated cousin, but once the beetle had adjusted its population swelled, and it became a first-class pest, moving eastward and resisting all attempts to curb its spread. To its would-be human exterminators, one of the more frustrating aspects of *decemlineata* ecology was its resistance to poisons; although potato beetles and the first insecticides appeared at the same time, poisons had little effect on the insect, which adapted itself quickly. During the 1870s, *decemlineata* spread to the eastern seaboard, which remained the insect's natural boundary until the First World War spread the insect through food shipments to France. The beetle was a major threat to French potato crops during the war, and between the wars, France's agricultural recovery was hampered by the continued presence of *decemlineata* in its northern departments. During the Second World War, the potato beetle crept eastward into Germany and Austria. At war's end, the pest had been identified as far east as Poland—a country at the border of the newly expanded Soviet empire.[34]

Potato blight and the voracious potato beetle were specific threats against which Quarantine Stations were designed to mobilize. In the lingo of the time, the stations would "organize for the struggle against agricultural enemies."[35] In spite of this, Quarantine Stations did not operate in a way that allowed them to deal effectively with any of the so-called enemies the state had identified. Disorganization, a lack of supplies, and a high percentage of food and crops that were not inspected all indicate that the work of the Quarantine Stations could have done little to limit biological invasions. On the other hand, as the stations grew and received better funding allocations from the Central Committee, they excelled at two jobs that were not specifically in their charter: surveillance over private plots, and discouraging individual production of potatoes.

The term "station" (in Russian, *stantsiia*) is a misleading one for the actual look of this new rural authority. Quarantine Stations were housed in poorly appointed offices in small towns near railroad lines. The most important and best-staffed Quarantine Stations in Ukraine were in the major cities of Kiev and Kharkov. The worst-staffed and most underfunded facilities were, ironically, on the Polish-Ukrainian border where the threat of cross-border contamination was most real. As with many postwar agencies, descriptions of work at the Quarantine Stations tended toward both propaganda and military jargon. In 1948, as the Quarantine Stations' authority was expanding (but before the stations had a chance to accomplish much actual work of inspection), stations were issued posters to display on their walls and kiosks with slogans in Russian and Ukrainian: "Colorado Potato Beetles: The Worst Enemy of Potatoes" and "Expose and Destroy the Potato Beetle." This rhetoric was borrowed from the bellicose language of the recent war, and indicates that while Soviet authorities shifted from human to animal enemies in the postwar period, they retained a strong sense that the Soviet countryside was still at war with potentially fearsome enemies.[36]

In the postwar era, the category of the *vrag naroda*—the enemy of the people—that had been a centerpiece of official rhetoric during the political purges of 1937 expanded to include nonhuman enemies. The implication was that beetles, foreign diseases, weeds, and other pests were new kinds of enemy agents for which Soviet rural citizens needed to be on the lookout. Soviet historians have noted that the notion of the omnipresent enemy was a classic theme of Stalinist propaganda in the 1930s and during the Second World War. In the words of Jeffrey Burds, "like *kulak, vrag* was a catch-all that utilized traditional concepts of ideological enemies to legitimize attacks targeting virtually anyone." This language, although often recognized as classically "Stalinist," was not unique to the Soviet Union under Stalin or to totalitarian states. Edmund Russell has persuasively argued that political and environmental conquests have often complemented each other in American history, and the bellicose propaganda that the United States employed after the First and Second World Wars against insect enemies capitalized on many of the same themes of exposing traitors; in fact, the work of waging a ground war and the work of eliminating agricultural pests were often accomplished using similar techniques and similar chemical poisons.[37]

In Kharkov Oblast, there were over a dozen different kinds of rural sites that the oblast's Quarantine Station was charged with inspecting in 1949 and 1950. The most numerous of these were kolkhozes, typically spatially organized around one or two villages. In 1949, there were also 380 individual (family-operated, noncollective) farms for the Quarantine Station to inspect. Other sites of inspection were the district's 69 state farms, 42 "potato seed distributing gardens," 36 "tomato seed producing operations," 4 scientific-experimental organizations, 37 grain-sorting points, 57 grain-collecting sites, 5 vodka distilleries, one grain elevator, one botanical garden, five indoor fruit houses, one airport, 97 train stations, and 214 "regional offices" (*otdelenii sviazii*).[38] These facilities all fall into one of three categories: points of transit, points of collection, and sites of distribution. Certain rural institutions, such as vodka distilleries, might fit into more than one category, but in general, Quarantine Stations were charged with inspecting food items in the ground before they were harvested, or at regional offices as they moved around the country.

Inspecting agricultural produce as it traveled made good sense, as plant diseases typically spread via foods bound for markets. Mimicking the practices of quarantine stations established in other countries before the war, Soviet stations planned to intercept contaminated food at state borders. However, there was a definitive shift between the small-scale projects of the very first postwar stations (1946–48) and projects carried out by the better-funded and better-staffed stations of the later (1949–51) period. Essentially, border control activities ceased, and although inspection of transit facilities continued, in the later period the real emphasis was on the stations' potato inspection departments, which acquired the largest number of personnel in these years. In order to understand the informal shift of focus that the Quarantine Stations had made by 1949, it is helpful to examine the day-to-day work culture of the facilities.

As was common with other Soviet enterprises of the time, an average of one-third of the people employed by the Quarantine Stations worked primarily at creating a paper record of the work of the stations. Throughout the period 1946–51, even the smallest stations employed a bookkeeper and a secretary in addition to research staff. Administrative jobs were challenging because typewriters and paper were scarce in the postwar era. Other jobs that station employees might hold included chauffeur (when the station was lucky enough to have been assigned a car), agronomist, *iadokhimikat* or poison specialist, "brigadiers," lab technicians, and lab assistants. Women held the majority of these positions, which was unsurprising in the male-scarce postwar countryside. Except for the positions of technical assistants and poison spreaders, which were

often short-term contract jobs, Quarantine Station workers
had a comparably high level of education. Most of them had
completed high school, and many had also finished one or two
years at a technical institute specializing in agriculture or pathology. This was impressive given that free public education
had been made widely available in the rural Soviet Union for
less than a single generation, and that the war had disrupted
many classes. It also indicates that as early as 1946, Soviet officials in the Ministry of Agriculture were taking care to hire
workers for their Quarantine Stations who had grown up or
had been educated away from the localities they were hired
to inspect, as most villages and small towns had only primary
schools. By hiring outsiders, the Ministry of Agriculture ensured loyalty from workers, although Communist Party members were scarce in Quarantine Stations.[39]

One such rank and file member who worked for the
Quarantine Station in the late 1940s was a woman named Lidiia
Mironenko, who held the position of "assistant poison-spreader"
at the Kharkov Oblast office. The station's offices produced reams
of paperwork, which is why a record of Mironenko's activities
at the station still exists today, but she mainly worked outside of
the office, on-site at various collective farms, individual small-
holdings, and regional offices. In the summer of 1948, she was
part of a team that spent two months gassing fields and barns
in a small village in Stanislav Oblast, a formerly Polish territory of Ukraine.

Mironenko's team used chloropicrin, DDT, and other
contemporary state-of-the-art pesticides, herbicides, and anti-
mycotics to treat diseased plants in fields and storage areas, and
to prevent outbreaks in places that the Quarantine Station determined were at risk. Although these chemicals were in short
supply after the war, larger Quarantine Stations had access

to them. The stations used the new poisons infrequently, but when they were used the stations often took photographs and issued press releases about their activities in order to highlight these sophisticated and relatively capital-intensive interventions. Although the countryside had only sporadic and unreliable access to machines and improvements like tractors, trucks, combine harvesters, hybrid seeds, and purebred animals, these artifacts received a disproportionate amount of good publicity. This helped contribute to a sense that the countryside was more highly industrialized and better-equipped than was truly the case. In the years after the war the number of foreigners allowed to travel in the rural parts of the country was limited. One U.S. Embassy official in Moscow forwarded to Washington a glossy picture book published in 1948, "not because they [the photos] depict anything outstanding, but because they are the first agricultural photographs which have been available for some time." The book contained pictures of combine harvesters at work, new grain elevators, and flax and sugar beet fields.[40]

In contrast to the scenes depicted in the government-approved book, Americans traveling by train through the Soviet Union reported seeing almost no machines in the fields during the harvest. One traveler reported that between Moscow and the southern city of Rostov, "at least 90% of all produce hauling was being done with horses and carts." Another traveler in 1948 reported that from a train between Moscow and Vladivostok (a journey of six thousand miles), "four combines were observed in the fields. Many horse-drawn reapers were observed but a good deal of the reaping was being done with scythes. Hand binding of sheaves was the rule." Except for travelers' reports, there are few records attesting to the scarcity of machines in the immediate postwar era.[41]

Bias could cut both ways: a Soviet official hosting a U.S. Agricultural Exchange Delegation in the 1950s complained: "In every kolkhoz . . . they visited, if some of the workers were barefoot, then the members of the group and especially the photo correspondents would take photographs at this moment. . . . Upon seeing the housing of kolkhozniks, MTS workers and sovkhozniks, members of the group, and in the first place the correspondents . . . were impetuously all the time photographing in an unfavorable way inside the workers' and MTS associates' apartments, and inside the kolkhozniks' houses."[42] Reports and photo albums such as the one created for Mironenko's group need to be interpreted with these biases in mind; the Soviet state was trying hard to create a record that placed machines and new technologies like field gassing at the center of a working, thriving Soviet industrial-agricultural landscape. Americans traveling through the countryside were inclined to focus on the opposite elements: bare feet and poor living conditions.

Although there were seven people in the Stanislav brigade, only two were permanent employees of the Quarantine Station: Mironenko and her supervisor, agronomist Gavriil Vakulin. Mironenko is a good example of a skilled middle-level worker in the Quarantine Station system. Born in 1926, she finished high school during the war and then took a series of short courses at a technical school to train her for work as a chemical technician. Her biographical data indicate that she had studied potato beetles for two years prior to accepting a position with the Quarantine Station. She was hired by the station in 1947, and although she was not a member of the Communist Party (very few Quarantine Station workers were until after 1950), she belonged to two state-approved professional organizations, the local Machine Tractor Station's Work-

ers Group, and the Zemorgan, or "Land Organization." These
memberships would have indicated to her superiors that she
was serious about advancing her career in agricultural science.

Four other members of the brigade included a scientist, a
technical assistant, a chauffeur, and a security watchman. The
group was outfitted with work uniforms, chloropicrin-spreading
canister injectors, and an extra daily ration of milk. The milk was
issued because the Kiev Institute of Hygiene recommended
that workers drink extra milk if they were working with chloropicrin to neutralize any poison they might accidentally in-
hale. The brigade also hired nine local men from the village in
Stanislav Oblast, who carried heavy equipment, dug holes, and
refilled chloropicrin canisters.

The stated task of the brigade was to "liquidate the isolated departmental hotbeds (*ochagi*) of potato blight" that had
been discovered by the State Quarantine Inspection the previous year. Their eight-week project required that they treat
five areas of the village—three at the local market, and two on
adjacent areas of private plots, which collectively comprised
937 hectares of land. In an early and unusual nod to worker
safety, the brigade held a workplace safety exercise before beginning their work with chloropicrin. First they hung a sign at
the edge of the field in which they were working, warning that
the chemicals they were using were dangerous. Three workers shared one gas mask, and before opening the gas canisters,
the group carried out a mock accident in which they simulated
their response to a gas leak in front of a small audience. Perhaps the presence of a photographer inspired this unusual
level of safety consciousness, because several dozen photographs of the work safety drills that were practiced exist now
in the archives of the Ministry of Agriculture.[43]

The brigade encountered problems in the course of its

work. Most of the difficulties involved the five canister injectors the brigade had been issued, which were supposed to be emptied by means of a small, kerosene-fueled pump, which was not easily available in postwar Ukraine. Additionally, the chloropicrin containers had been shipped to the work site in 380-kilogram barrels that were almost impossible to move, especially given the shortage of draft power in the villages being treated. The villagers' small horses, which the brigade conscripted for work, struggled to drag the barrels across the fields, and in photographs the barrels dwarf the animals. The barrels arrived with no obvious openings, and brigade members had to fashion openers out of used rifle cartridge clips to open the cans.

Treating potato blight in Stanislav Oblast was expensive for the state; the 1,887 kilos of chloropicrin the station used cost 17,756 rubles.[44] This was approximately the cost of a new car, an asset that the Quarantine Station did not possess. The high cost of such a project had little or no relationship to the economic value of the farms being treated, or to the level of risk posed by the blight itself, especially as it was never determined whether the blight in Stanislav was an outbreak of a milder, indigenous strain of the local blight fungus, or whether it was truly the same disease that had contributed to famine in Ireland a century earlier. Instead this exercise and others like it were primarily displays of power and state authority. By placing a high price on such chemicals the state made them important, but in doing so it made them too expensive for collective farm organizations.

At least the potato blight existed. With the exception of a few suspected (but never corroborated) sightings of potato beetles in L'vov Oblast (another formerly Polish territory), there was no evidence that the Quarantine Stations in Ukraine encountered potato beetles anywhere in the region, in spite of intensive searching. The first official paperwork that began the hunt for potato beetles was issued on May 28 1948, and called for Quarantine Stations to begin inspecting for potato beetles already in June. Potato beetles were to be hunted on all 1,726,000 hectares of Ukrainian territory. Schoolchildren and Pioneers were encouraged to spend Saturdays roaming potato fields, and many Quarantine Stations planted experimental fields as bait to lure the beetles out of hiding. One important new activity of Quarantine Stations after 1948 was to host public education events that would explain the threat of the potato beetle to kolkhozniks and private producers who raised potatoes, presumably encouraging these farmers to be on the lookout for the pest, and also perhaps to consider planting other crops as a way to avoid the impending threat of beetle infestation. The Ministry of Agriculture provided films, full-page newspaper ads, lectures, and posters as informational propaganda that the Quarantine Stations could distribute. In spite of these efforts, potato beetles either stayed extraordinarily well-hidden in Ukraine between 1948 and 1951, or they were not present at all.

Between 1946 and 1952, the Soviet Union's agricultural production, although low, was not much damaged by insects or blight. Colorado beetles wreaked havoc with potato crops in Western Europe immediately after the Second World War, but they did not become endemic to the Soviet Union until the early 1950s. Likewise, the cases of potato blight that Ukraine experienced before this period were few and far between. Given these facts it might be easy to conclude that the Soviet Union's Quarantine Stations were effective at controlling these biological invasions and that the work they did in the years after the war helped to contribute to the recovery effort and the stabilization of the rural food supply. However, by looking more closely at the ecology of these diseases and at the kinds of work Quarantine Stations performed, it becomes clear that the stations did little to prevent or control biological invasions. The perceived and publicized threats from blight and beetles in the immediate postwar era were likely exaggerated or just plain false. Because at least one strain of potato blight was indigenous to Ukraine, there is no proof that the blight the Quarantine Stations combated in 1948–50 actually posed a severe threat in the region. At its peak, potato blight only ever affected seven of twenty-five oblasts in Ukraine. Imports and exports remained a potential vector for disease, but there is no evidence that blights, beetles, or other scourges were actually transmitted in this way in the postwar Soviet Union. Between 1946 and 1950 only one infectious disease was identified in shipments of produce arriving in Kharkov.[45] Although a certain level of preparedness in combating such invasions is generally warranted, the response level from the Quarantine Stations in the later 1940s was out of proportion to the threat that the countryside faced from agricultural pests.

Other threats to agricultural productivity in Ukraine, such as weed infestations or low germination yields from seed grain, were ignored. These challenges hindered agriculture far more than any pests and might have been more appropriate problems for the Quarantine Station to address. The Soviet government neglected the low germination rate because it would have meant admitting that the seed it provided was substandard. The state ultimately addressed the weed problem at the kolkhoz level, but its initial interventions were propaganda-based and ineffective. If the Quarantine Stations did not address the most basic and pressing issues of contamination that affected postwar agriculture, and if they were not preventing disease outbreaks in the postwar period, then what did they accomplish? Why did they receive so much money from the Ministry of Agriculture?

To begin with, it is important to note that the imagined influx of potato beetles and various plant diseases was also an aftereffect of war. Beetles, viruses, and other biological infestations were a new kind of hidden enemy for the state to focus on after the threat of military invasion had passed. These two types of invasions, military and biological, were linked for the Soviet state, and also, perhaps, in the minds of Soviet citizens. Germany or other hostile powers still posed a potential hidden threat to the Soviet countryside, allowing biological invaders to sabotage the food supply, and thus the political stability of the Soviet Union. Certainly this was true just a few years later during the Cold War. In East Germany a propaganda pamphlet accused American planes of dropping potato beetles on a village near Zwickau, and called on the Americans to cease and desist from such hostile actions. The pamphlet explained, "Colorado beetles are smaller than atomic bombs, but they are also a weapon of U.S. imperialism against the peace-loving working population."[46]

Because of increased funding from the Ministry of Agriculture in 1948, the Quarantine Station program had swelled into a major operation by 1950, almost quadrupling the size of its staff in the three years between 1948 and 1951. The largest and fastest-growing department was the one that dealt with the control of potato diseases. Although the department employed laboratory scientists and even had a propaganda branch, most employees did not treat or study potato diseases; instead they focused on collecting records of who was growing how many potatoes. Although the stated purpose of collecting such records was so that Quarantine Stations could better control blights and beetles, the secondary, unstated purposes of this data were probably far more useful to the state. By acquiring an accurate count of potatoes and potato acreage, the Quarantine Stations helped to establish accurate tax records, and the fact that the stations could potentially help increase state revenue was not lost on the station inspectors. Allotment gardens were taxed in several different ways in the postwar period, but the highest taxes were levied on foods produced for private markets. Because potatoes grow and are stored underground, it was a challenge for the state to keep track of growing levels. By sending Quarantine Station experts out into the field and into storage cellars, the state was able to gather better records. Even more important than acquiring better data was the fact that the Quarantine Station could show collective farmers that the state had the power to collect this data.[47]

Between 1948 and 1950, the number of potatoes produced on allotment gardens in Kharkov Oblast fell by half, from over 50,000 to 23,000 hectares. The number of private plots also fell, from 380 in 1949 to 65 in 1950. Similar reductions were recorded in other regions. This decrease has been explained in a variety of ways: the rise of a harsher system of rural taxation, the decline of village markets, an increase in the level of control collective farm chairmen had over their workers. These interpretations are all correct, but historians have overlooked the mechanism behind these changes. Directly and indirectly, the Quarantine Stations were foundational to increasing state control on collective farms. The stations had the authority to seize privately grown produce in the name of food security, as well as the authority to prevent markets from being held. By accurately reporting how much land households planted in potatoes, the quarantine authority made that land available to taxation by the collective farms. Inspection of private markets gave the state a way in to examining the informal economy of individual production and establishing an accurate tax base. Kaplan has argued that the outsider status and relatively high level of education that new collective farm chairmen possessed in the early 1950s was a major mechanism for rural change, but the first arena in which highly educated outsiders were employed by the Ministry of Agriculture in rural areas was the Quarantine Stations, not the collective farms directly.[48]

After a war, and especially after an enemy occupation, a return to peace is always preceded by a period of adjustment and change. In the case of rural Ukraine, a major theme of this transitional period was food scarcity and insecurity, which resulted in famine in parts of the country and the decision on the part of many peasant farmers to continue raising the majority of their own food in the form of potatoes. During the war, potatoes grown on allotments had been outside the purview of the Ministry of Agriculture, but the work of the Quarantine Stations in the postwar era brought them back under state scrutiny. By publicly establishing that the Ministry of Agriculture was invested in monitoring this sector of the informal economy, Quarantine Stations encouraged peasant workers to return to state-approved work—above all, growing grain on collective farms. Collective farm work did not confer the benefits of allotment gardening, but on the other hand there was no direct financial penalty involved with such work, such as the high taxes that were imposed on marketing potatoes by the 1948–49 growing season.

In some ways, Quarantine Stations resembled another famous Soviet watchdog agency, the Machine Tractor Station or MTS system. The differences between the two organizations highlight the strengths of the smaller, newer Quarantine Station system, which is essentially the strength of a focused agency over a large and diffuse one. Unlike the MTS, the Quarantine Stations did not attempt to acquire an all-encompassing view of the countryside. Stations were wherever the Ministry of Agriculture had identified problematic border crossings and regional offices. The long Ukrainian border with Poland received a significant number of these stations. Quarantine Stations were not engaged in an important primary task such as operating a national fleet of farm machines like the MTS, which meant that these institutions had resources to devote to small-scale, intensive surveys, pest treatment programs, and on-the-ground interactions with locals. Although the Quarantine Stations worked to eradicate pests and diseases from fields, devastating outside infestations did not arrive in the immediate postwar period; instead, internal threats like weeds and poor work organization posed the largest obstacles to agricultural productivity.

The work of anticipating biological invasions and that of reestablishing a tax base and a present state authority in the countryside by counting potatoes, locating root cellars, and raiding local markets were in fact related tasks. While the larger and more impersonal Machine Tractor Stations kept watch over kolkhoz activities and reported irregularities to the central authorities, Quarantine Stations were more selective in what they chose to examine; their initial targets were simply the large potato crops produced on private allotments. Because these were private, the kolkhoz administration and the MTS would not normally have had access to them, but in the interest of preventing the spread of diseases and pests, Quarantine Stations did have such access. Biological invasions did not discriminate between public and private places, and so Quarantine Stations did not do so either. While the MTS observed and reported on the work of individual kolkhozes, Quarantine Stations could focus on units as small as allotment gardens, sharing this individualized information with tax authorities. Potato growing was an undesirable behavior in the eyes of the state, but in the postwar period the best way to combat such behavior was through economic disincentives rather than the strong-armed shaming efforts and disenfranchisement for which the MTS had become famous.

The decrease in potato production after 1949 was not entirely due to Quarantine Station intervention, but the kind of work the stations performed in the postwar years marked an important and enduring change in the way the Soviet countryside was governed. By the early 1950s, other entities such as the state procurement agency and the collective farm administrations had begun to acquire more power in rural areas, using some of the same methods of intense scrutiny in the name of improvement and reconstruction. The Quarantine Stations are just the earliest example of a new kind of regime of rural control that endured throughout the 1950s. The fact that Ukrainian peasants chose not to grow potatoes in spite of the fact they were still a potentially life-saving and profit-making crop would seem to indicate the presence of a rural authority in the countryside that was capable of translating the state's desires into a language of observation and control that peasants could understand.

3

Animal Farms

In March of 1953, after almost thirty years in power, Joseph Stalin unexpectedly died. After a brief struggle for power, Nikita Khrushchev emerged as Stalin's successor. The new leader adopted a very different approach to improving the Soviet countryside and increasing agricultural production. Unlike Stalin, who visited only a handful of farms and villages during his time as leader, Khrushchev had been born and raised in the Soviet countryside; he was an agricultural insider. Even before Khrushchev became the leader of the Soviet Union he focused on agricultural modernization in his native Ukraine, where Stalin had appointed him premier of the Ukrainian Republic during the Second World War, and afterward until he resigned to return to Moscow at the end of 1949. During the decade Khrushchev spent at the head of the Soviet Union, his leadership was characterized by reform and the return of a civil society not governed by terror. Just as remarkable as Khrushchev's efforts to de-Stalinize the country was his enthusiasm to reform and modernize Soviet agriculture.

One of Khrushchev's most memorable reforms was also

one of his least practical: in a speech in 1956, Khrushchev committed the Soviet Union to catching up to and surpassing the
United States in meat and milk production. For a country that
had always produced a fraction of the U.S. output of meat and
dairy products, this was a bold promise. How did the Soviet
state intend to increase its supply of these food items? Why
was such a goal important to the new, post-Stalinist government? Just as with the policy of biological control in kolkhoz
agriculture after the war, raising meat and milk production
was intended to accomplish several goals simultaneously. In addition to providing Soviet citizens with more and better food,
the campaigns to increase livestock populations were popular
public forums that showed off the Soviet state's ability to overcome long-standing environmental and biological barriers to
agricultural productivity.

The Ministry of Agriculture emphasized how techno-
logically sophisticated and well-ordered its animal agriculture
was. Through careful application of science and technology,
Soviet farm animals exhibited the Soviet Union's mastery of
modern agriculture. In fact, the methods behind state initiatives to increase livestock populations were often quite basic:
better food, better housing, and faster breeding cycles for all types
of farm animals. Due to postwar scarcities, the Soviet Union did
not have the machines, advanced feed regimes, prophylactic antibiotics, and other inputs associated with the industrialization
of animal agriculture in North America and Western Europe.
Nonetheless, this sector of the USSR's economy was rapidly modernized and animal production was reordered, intensified, and
dramatically improved.

Much of this progress was credited to Trofim Lysenko, a
scientist whose hostile attitude toward genetics might seem at
odds with the improvement of livestock. Lysenkoism has become a shorthand way of describing misleading, ideologically

driven science. Surprisingly, in the context of Soviet animal agriculture, Lysenko's recommendations helped animal farms become better-organized and more productive. As a theory of human and animal management, Lysenkoist ideology was a useful tool. The basic tenets of Lysenkoism were simple to understand and they played to the Soviet Union's postwar strengths by emphasizing human labor and by drawing on long-standing Soviet projects of acclimatization and hybridization. Lysenkoist practices also introduced an element of showmanship that had been missing from the humble sector of animal agriculture. Khrushchev's government was genuinely concerned with producing more and better meat and milk products, but it correctly perceived that its planned improvements could also demonstrate how a new, socialist approach to animal agriculture was superior to capitalist methods.

Improving Soviet animals posed special challenges to the Soviet state. The slow war recovery, an unusually cold climate, and the genetic isolation of farm animal populations all hindered productivity. The Second World War had been devastating for animals as well as humans; over half the population of horses and draft cattle, a third of all cows, and half of all pigs were killed.[1] At the time of Stalin's death in 1953 these numbers had only just recovered to their prewar levels. In addition to the draft shortages discussed in the second chapter, animal products such as meat, milk, and leather remained in short supply well into the 1950s.

Livestock populations also recovered slowly because the war disrupted breeding cycles and destroyed breed records, and war and postwar chaos spread diseases among animals as well as people. In the immediate postwar period the Soviet state was concerned about communicable animal diseases in addition to the those of plants. In 1945, on collective farms in Kiev, Kharkov, and Leningrad Oblasts, for example, chickens and pigs were placed under quarantine to protect against an outbreak of plague. In 1946 in Novosibirsk animal health and housing had been so damaged by the war that the kolkhoz of the city's meat trust wrote to the central government to ask for an extra animal technician and funds for new pig and cow barns in order to help safeguard against infectious diseases.[2]

Fighting animal illnesses remained a major theme throughout the postwar recovery. From September through the end of spring, rural newspapers reported extensively on the health of animals, emphasizing the preventative measures that would prevent scarlet fever, rickets, plague, and severe malnourishment. Extra workers were hired to clear manure out of barns and to take pigs and cows for brief walks. Due to a shortage of fully accredited veterinarians, animal technicians and veterinary assistants typically oversaw animal welfare, often in newly created field stations that specialized in the community health of agricultural animals.[3]

A major risk factor for animals during the winter was food scarcity. High-quality feeds were in short supply for much of the late 1940s, and even marginal feeds like potatoes and turnips were earmarked for hungry humans and not animals for most of that decade. As late as 1950 many collective farms were only able to amass half of the silage they anticipated needing for the winter season. Many animals that had survived the war starved during the winter in the years immediately following because farms and private owners could not amass enough food to maintain them for the long winter months. This was not just a problem of the postwar years; the specter of winter food scarcity loomed large throughout Russian history. Until the postwar period large animals such as cows, sheep, and pigs had depended on two very different seasonal feeding regimes. In the summer, animals fed themselves without much human assistance. In fact, by consuming foods like grasses, shrubs, kitchen scraps, and fallen acorns, domesticated cows, sheep, and pigs spent this part of the year converting foods that were inedible for people into high-quality protein sources, thus ultimately enriching human diets. Summer weather also favored procreation. Although most domesticated animals could breed year-round, more offspring survived in warm weather. In winter weather animals relied on humans for food, and this took far more trouble and energy than the summer regime. In the winter animals grew more slowly, and slowed or stopped producing milk.

In the Soviet Union, larger farm animals functioned as draft labor, a source of food, and a source of materials such as leather, bones, and wool. But animals also took a lot of work and trouble to own and maintain, mainly because of the cold environment. The time and cost of feeding and housing animals during the winter could be prohibitively expensive. Even the warmest parts of the country had a colder climate than most European or North American averages and this meant that feeding animals over the winter was a significant liability for farmers that left little room for waste or excess.[4] Historically, Russian households had little use or desire for the burden of large animal stocks, and farm animal populations in the Soviet Union were low and not as productive as livestock in Western Europe or North America. Historians have often interpreted the relative scarcity and low productivity of Soviet animals as proof of poverty, rural hardship, and state ineffectiveness. Nevertheless, both state and citizens adapted to these circumstances. As we will see in the next chapter, both traditional cuisines in Russia and new Soviet mass-produced food items capitalized on the few meat and milk products that were available, and alternative consumer goods like wooden and glass buttons and cloth and rubber shoes were common replacements for scarce leather and bone products.

Well into the postwar period, large farm animals were absolutely crucial for one aspect of agricultural production— draft labor. Draft animals rather than machines provided the bulk of field horsepower for smaller collective farms across the Soviet Union well into the late 1940s. Harvests on these farms were smaller when there were draft labor shortages, and this in turn fueled a vicious cycle because kolkhozes raised grain and root crops primarily for use as animal feeds, while grain for humans was grown on the much larger and more highly mechanized sovkhozes. Thus a shortage of draft animals turned into a shortage of feed crops, forcing kolkhozes to either reduce animal populations or purchase expensive animal feeds from the state.

In the postwar period, the Soviet Union prioritized restoring tractors to their prewar numbers, a goal the Ministry of Agriculture claimed to have accomplished by 1950. In part, this success was thanks to a new way of measuring tractor power. The Fourth Five Year Plan listed a loss of 137,000 tractors in war, and a goal of producing "no less than 325,000 tractors." By 1947, when the revised Three Year Plan for agriculture was adopted, this goal had shifted and tractor production more often described in terms of replacing horsepower and of "doubling" or "tripling" the number of tractors that were being produced each year, than of actual figures. New plants began producing a smaller number of higher-horsepower tractors, thus making it easier for the state to reach the vague targets set forth in the revised plan.[5]

Because Soviet animals followed a seasonal eating regime and were used as draft labor, their evolution as industrial organisms differed from the path their Western counterparts had taken. The industrialization of animals in the United States and the United Kingdom first eliminated the seasonality of breeding and eating patterns, and then uncoupled the meat and milk industries from the animal by-products industry as the plastics and plant oil industries expanded.[6] In the Soviet Union seasonality endured and the plastics and plant oil industries lagged, thus animal industrialization followed a different course from capitalist countries. Postwar animal farms were standardized and production was intensified but the farms retained some premodern characteristics, notably the seasonality of eating and breeding regimes. They also relied on human caretakers to oversee the health of animals, and they continued to value animal by-products like hides, bones, feathers, and fat that remained important for making clothing, home goods, and cooking oils.

Before the war, state policies for increasing animal populations varied greatly in their effectiveness. Immediately afterward, there were fewer animals and state policies were even less clear. The Ministry of Agriculture refused to recognize that the majority of animals that had survived the war were privately owned, and the government continued to impose, and in fact increased, taxes on privately owned animals in order to discourage individual farmers from increasing stocks. Members of a kolkhoz were permitted to own either a single cow or three sheep or goats, and one pig. They paid taxes on the animals they owned and an acquisition fee if they purchased a new animal or if their livestock reproduced.[7] If individuals owned more stock than the legal limit, they had to pay a steep fine. Kolkhoz members did not always own the products even of animals that they owned free and clear. For example, once cows and goats began to produce milk, the local dairy trust requisitioned a large percentage of it, in some cases over half the output. Thus, there was little incentive for collective or individual farmers to breed the animals they owned or to otherwise build up their stocks.

Animals that belonged to kolkhozes were typically worse off than their privately owned counterparts as regards both food supply and caretaking. Foreign observers reported that collectively owned animals looked scraggly and thin, and that only privately owned milk cows looked as if they were capable of giving significant amounts of milk. Although newspaper reports claimed that stocks were increasing rapidly in the immediate postwar period, statistics from this era do not bear out this claim. From an economic perspective, there were few incentives for survival-minded peasants to keep collectively owned animals alive and healthy, as their meat and milk products were largely purchased at a loss by the State Provisioning Board and various state-operated trusts and co-ops.[8]

A further challenge for animal breeding was the small population of purebred stock animals. The Soviet Union had purchased purebred animals from the United States and various countries in Western Europe before the war, in order to stock the early model animal farms created after the first wave of collectivization, but these animals only made up a tiny percentage of the total livestock population in the postwar Soviet Union. Stable purebred and hybrid animal lines had been abandoned during the war, and stud books and other records had been lost.[9] Typical Soviet farm animals were small, sickly, and slow-growing—the exact opposite traits from what is desirable in a population slated for rapid industrial expansion. In the early 1950s the Soviet Union still lacked basic resources such as the heated indoor housing and high protein feeds that George Heikens and Guy Bush had requested for their farms in 1930.

Even so, the heterogeneity of local animal populations had advantages as well as liabilities. Soviet animals were often more cold-hardy than breeds that had been developed in temperate climates. Cows and sheep had dense insulating coats, and their hides, used for winter clothing, were important by-products for the private farm sector and a secondary source of income for independent farmers. In pigs, slow growth translated into thick insulating layers of edible body fat. This fat, *salo,* was a culinary staple in many parts of the USSR. The small size of Soviet animals was undesirable in the eyes of agricultural modernizers, but compactness was historically a positive trait. Small animals ate less and conserved body heat better, and thus were better suited for over-wintering in a harsh environment with sparse food and little heat, such as that found on most Soviet kolkhozes. The premodern look, size, and behavior of Soviet animals may have been detriments to agricultural planners, but they were all clever adaptations to the cold environment and scarce resource base in which these animals had developed.

The decade after the war was also a time to improve, rationalize, and reorganize collectivized animal husbandry. American visitors to the Soviet Union after the war often remarked on the appearance of farm animals they saw as they traveled the country, comparing Soviet stocks to the familiar creatures of their home country. Soviet agricultural specialists were rarely candid about the state of the animal population, so these views from outsiders offer a valuable perspective. On the one hand, the Soviet Union maintained a number of model farms stocked with sleek, plump, purebred, prize-winning animals, and comments about these operations were uniformly favorable.[10] At the other end of the spectrum were the animals Americans spotted from the windows of cars or trains. These more typical

Soviet specimens elicited much less admiration. In diaries, reports, and letters home, travelers disparaged this population, noting generally how small, dirty, and feral they appeared.

One reason such animals were visible at all was that they were roaming outdoors close to train tracks and roads. This free-ranging lifestyle was appropriate for beef cattle, but by 1950, pigs and milk cows living outside looked out of place to Americans. Such animals would have been especially obvious due to the lack of fencing along the railway corridors, which commonly resulted in herds of cows and sheep straying onto the tracks, forcing trains to slow down or stop in order to avoid collisions. Railroad corridors were maintained in a way that encouraged the growth of short grasses, and as a result animals and their guardians saw railway right-of-ways as choice spots for free grazing. The state imposed a hundred-ruble-per-day fine on owners and kolkhozes that allowed their animals to graze near unfenced right-of-ways, but there is little indication that these fines were effective at clearing the tracks.[11]

American reports reveal as much about the advanced and highly specialized state of American animal husbandry in the postwar era as they do about the status of Soviet animal farms. One American traveling by train in 1948 noted that "herds of 40 to 65 cattle are quite common in the central Ukraine. All herds seen were of mixed breeds . . . none of the cattle observed could be classified in a beef category and no udders were seen that could honestly be judged as belonging to a 2500 liter a year cow. The general flesh and hide condition appeared good [but] general barn sanitation can be deduced as poor; most cows' flanks and bellies were thickly caked with dung." Another traveler remarked that in Ukraine, "One traveled for miles only seeing one herd every few miles. . . . herds usually numbered about fifty, composed of a motley mixture of cattle of different colors . . . well nourished but often scrubby. They were practically all dairy types, to the extent that they could be typed at all. No heavy set beef cattle were seen." In the United States, cattle had diverged before the Second World War into "dairy" and "beef" cattle. Beef cattle had undergone the more dramatic metamorphosis, as breeders selected to maximize the amount of meat on a carcass. By the postwar period, American beef cattle were enormous and muscle-bound. Dairy cows retained their earlier appearance, although their udders grew, as twenty-five hundred liters became a standard annual milk output per cow. The motley herds American observers noted in the Soviet Union would have been unfamiliar because most American herds in the postwar era were composed of just a single, specialized, purebred animal variety.[12]

In 1951 one traveler noted that "all herds observed appeared thin and showed all signs of having had a tough winter. Privately owned cattle were in much better condition [than collectively owned cattle]." He also noted the mixed herds, stating that "the same motley type of cattle as observed in all other parts of the Soviet Union were in evidence." Other American visitors to the USSR between 1948 and 1951 noted the diminutive size of Soviet cows, the obesity of Soviet pigs, and the fat and meaty tail of the typical Soviet sheep—a characteristic derived from Central Asian breeds that most foreigners had never seen.[13]

While these chance sightings provide a credible account of the state of postwar livestock, they remain unofficial reports. Visitors were supposed to view Soviet livestock in model collective farms that had been created in part to impress foreigners.[14] The well-kept animals on display at these farms did not give an accurate impression of the national state of animal affairs, but they could reveal how the Ministry of Agriculture prioritized certain qualities above others in livestock. Until 1955, animal productivity was a primary concern; this was measured in terms of fecundity, weight gain, and milk production. The better-looking animals on display at model collective farms and agricultural exhibitions were compared favorably to well-known purebred animals by American observers. Visitors noted that putatively original Soviet-created breeds of cows and pigs on exhibit in model farms sometimes looked exactly like purebred animals from other countries. The similarities between the Baltic Brown and the Swiss Red cow, the Khalmagar and the Holstein cow, or the Lithuanian Bacon and the Yorkshire pig were often too striking to go unremarked in travel diaries. It was a fact that the Soviet Union had copied American, British, and German designs for automobiles, airplanes, tractors, and combine harvesters. Visitors to Soviet farms often suspected the Soviet Union had also reverse-engineered its animal breeds.[15]

The discrepancies between the animals on display on the Soviet Union's model farms and those glimpsed from train windows constitute an example of the largest ongoing challenge Soviet agricultural professionals faced in rural industrialization: that of making modernity universal and bringing the countryside up to speed with innovative practices, materials, and the latest improvements in animal agriculture. The difference in livestock quality between the model farms and the "motley" animals living on the majority of kolkhozes was a pattern repeated in all areas of agriculture. The Ministry of Agriculture could certainly create and provide models of systems that worked well, such as its network of agricultural experiment and animal breeding stations, but it could not successfully disseminate these models and make them the standard for the whole country.

There were many reasons why the Soviet Union was unsuccessful at diffusing livestock modernization beyond its model farms. Rural workers usually had a low level of education and skilled workers with advanced training were reluctant to relocate to rural areas. The roads between town and country were poorly maintained and the sheer size and diversity of the territory was daunting. Finally, the lingering vestiges of estate-based agriculture rewarded a cautious and traditional approach to farming. The Ministry of Agriculture widened the gulf between modernized model farms and the more widespread traditional-style kolkhozes by giving better funding to successful and profitable kolkhozes. Those that lost money rarely received new equipment or choice stock; their very unprofitability made them appear backward to agricultural authorities within the Ministry of Agriculture. This practice formed a vicious circle where positive publicity, foreign tourist groups, and high-quality animal stock were all funneled into a small number of well-working collective and state farms, while the majority of farms remained neglected and in debt. Thus, farms that had achieved early success continued to be successful and well-supported, and farms that struggled were largely left to their own devices by the state, which did not wish to associate too closely with failing enterprises.

The slow dissemination of modern improvements inspired the Ministry of Agriculture to develop new policies in the postwar period to end the trend of uneven progress; its goal was to reorganize patterns of work across the kolkhoz system. Although uneven progress was rarely mentioned in the Soviet press or by the government, the Ministry of Agriculture worried privately about how best to modernize the parts of the Soviet countryside that were at risk of getting left behind. This was especially true after Khrushchev came to power. In the 1930s, tractors and electric lights arrived in the countryside as two symbolic material components of modernization, but in the 1950s the state's approach was different. Rural modernization was now a problem of labor management for the Ministry of Agriculture. In the eyes of the state, unsuccessful farms lacked a rational approach to the organization of their work and good management became more important than good materials. Throughout the 1950s, policies on kolkhozes reflected this new value. Postwar agricultural policies demanded that kolkhoz members adopt a new approach to organizing the work of farming instead of adding machines, electric lights, or new chemicals to the mix.[16]

To this end, the Ministry of Agriculture focused on reforming rural mentalities and sclerotic rural work patterns. The ministry moved to consolidate collective farms between 1946 and 1951, it shifted kolkhozes to brigade-style agriculture beginning as early as 1946, and introduced new, simplified land use patterns on collective farms.[17] Although postwar agricultural reform is most closely associated with Nikita Khrushchev, these three key changes were all initiated well before he came to power in 1956. These land reforms and farm reorganization campaigns were intended to improve management and labor efficiency on smaller, underperforming kolkhozes where losses were widespread.

Similar to campaigns that made both labor and land use more efficient, reforms targeting animal farms were intended to make these sites more productive and to increase the efficiency of agricultural workers. To this end, the state created and reinforced new management policies that encouraged caretakers to focus on animals. The theories of Trofim Lysenko were fundamental to these reforms. Although not originally focused on animal agriculture, Lysenko's ideas were central to the postwar effort to recast the work of animal agriculture as a process that could be controlled through scientific management. This was not the first time a form of scientific management sought to reorder the countryside, but it was the first time such a measure was successful.[18] Lysenkoism allowed the Soviet state to mobilize both its human and its nonhuman agricultural resources as agents of political change; cows, pigs, and sheep were not simply multiplying and growing fatter, they were improving in ways that conformed to socialist (and Lysenkoist) ideas of progress and evolution. Kolkhoz workers who became swineherds and milkmaids were not simply accepting new jobs; they were expanding a new category of rural worker whose identity (female, skilled, capable of keeping written records) was expedient for the postwar Soviet state.

Soviet labor planners first tried to impose scientific management in the 1930s when they embraced Taylorism—the quintessential American system of scientific management—as a practice that was relevant to socialist factory work. Taylorism, developed under a capitalist system in order to maximize the profitability of worker time, was adopted by the Soviet state as a capitalist technique that could be exploited by a communist society. Fordist and Taylorist principles and techniques such as time and motion studies were essential to the reorganization of factory work.[19] But in the 1930s, scientific management made few inroads in the rural Soviet Union. The fields of Soviet farms in the 1930s rarely resembled factories, and industrial systems of agricultural production broke down more often than they succeeded. American-imported processes that increased production and efficiency in steel making, machine building, and light industry failed to be adopted on Soviet collective farms during the 1930s. Soviet propaganda trumpeted a high level of mechanization and industrialization on the farms of the 1930s, but well into the 1950s humans and animals, not machines, were the main sources of labor power for collective farms. While there were a few large-scale grain sovkhozes that proved mechanization worked sometimes, the 1930s Soviet countryside was simply not capable of being organized as an industrial system on a mass scale.

At first glance, the imprecise, politically motivated science of Lysenkoism and the hyperrational system of Taylorism have little in common, but a closer look at the evolution of Lysenkoism reveals that it was popular and enduring because it could be co-opted by the state as an organizational approach to scientific farming. This is similar to the way that the capitalist ideologies of Taylorism and Fordism were adopted and improved upon by the Soviet state to serve as the basis for a socialist push toward industrialization during the 1920s and 1930s.

At the beginning of his career, Lysenko developed and promoted his scientific ideas in a sincere effort to improve cereal yields, but by the 1930s Lysenko's political ambitions had altered the way in which he presented his ideas publicly. Like many successful scientists of the time, he reworded his arguments and used the language of Marxism in order to curry favor with his superiors.[20] By 1949, the science of Lysenkoism had become an official and well-publicized state policy. Once this occurred, Lysenko himself had little control over the ways in which his ideas or words were used to further state goals. Excerpts from his speeches and writings were often reprinted in newspapers and in agricultural periodicals, but these reprints were selective. Just as Lysenko had modified his own ideas to fit the popular Marxist socialist rhetoric, the Ministry of Agriculture chose to emphasize certain aspects of Lysenkoism and ignore others.[21] Thus a theory of heredity and a scientific practice, twice distorted, became a platform for the state to further its own goals of scientific management in the countryside.

Lysenkoism became official state policy in 1949, but Lysenko had been creating a body of scientific research, advancing himself professionally, and publishing papers and essays that combined his scientific research with politically expedient ideology for over two decades. Lysenko's research appealed to a scientifically oriented state that was interested in improving itself and in defining a scientific agenda that would set it apart from Western, capitalist models of science. Lysenko's theories on plant and animal breeding, while nominally based on Darwinian evolution, also accepted some of Lamarck's theories —discarded by most scientists in the late nineteenth century— that acquired characteristics might be at least partially inherited, and that plants and presumably animals could improve their genetic stock if they were provided with the proper environment. He believed that by either environmentally challenging or nurturing an organism, one could "shatter" its hereditary tendencies and refashion them. Geneticists had discarded the notion that the environment could influence inheritance for most of the twentieth century, and Lysenkoism was widely derided in Western publications for adhering to these theories.

The idea that genetic stock might improve when it encountered challenging environments was appealing in a country where the climate and soil quality had already played a fundamental role in determining the scale and success of agricultural enterprise. Opposing Mendelian (heredity-based) genetics was also a shrewd political move on the part of Lysenko. Gregor Mendel had been dismissed by the Soviet state as a bourgeois (and therefore corrupt) scientist in part because his research was based on laboratory experiments rather than field studies. Soviet genetics labs, like their American counterparts, had long since separated from their practical origins in nineteenth-century programs of agricultural breeding. In contrast, Lysenkoism never lost its farming interest. In the words of Lysenko, "close contact between science and the practice of collective farms and State farms [enables] us to learn ever more and more about the nature of living bodies and the soil." Lysenko called his system of improving plants and animals "agrobiology," a name that highlighted the close connection between agriculture and scientific research.[22]

Lysenko claimed there were simple, universal principles governing the development of all living organisms and this caused him to make sweeping statements about many aspects of biology, but his direct research experience was with the agricultural crops of wheat, rye, and potatoes. In his early career he achieved national recognition for groundbreaking experiments on vernalization, or the cold hardening of wheat, in Azerbaijan in the 1920s and 30s. Lysenko also increased potato yields in cold parts of the country by modifying planting schedules. Lysenko's non-Mendelian theory of heredity led him to publicly attack the emerging scientific discipline of genetics, as well as one of its most acclaimed practitioners, the biologist Nikolai Vavilov. The rivalry between Vavilov's belief in genetics and Lysenko's insistence on agrobiology persisted throughout the 1930s. Lysenko gained a definitive upper hand when he was appointed director of the All-Union Agricultural Academy in 1938. With this appointment, Lysenko gained the highest position of power at the largest and most prestigious agricultural research institute in the country. Vavilov served under Lysenko at the academy for several years, but ultimately his work in the increasingly unpopular field of experimental (rather than applied) genetics lost him the favor of the Stalinist regime, and he died during the Second World War in a Siberian prison.[23]

By contrast, in the face of mounting evidence against his research, Lysenko remained a popular and well-supported figure until 1965. During this long era he wielded significant influence over the direction of agricultural research and the general focus of agricultural policy. Historians have used the waxing and waning of the careers of Lysenko and Vavilov as an allegory of the contamination of Soviet science by socialist politics. At various times in Soviet history the intellectual integrity of linguistics, psychology, chemistry, and cybernetics were similarly compromised by the confluence of ideology and science. However, a historical interpretation that reduces the experience of Lysenkoism to allegory is limited.[24] While there is no question that Lysenkoism represents a troubled historical confluence of science and politics, this was not the sole reason for its success and endurance over a period of nearly thirty years. Lysenkoism was also functional and relevant to the needs and abilities of Soviet collective farms in the state's quest to improve these facilities in the decade after the Second World War.

For most of his years in power, Lysenko limited his scientific publications to the agricultural subject he knew best —economically important cereals—but in his speeches and popular publications he tended to write and speak more broadly about both "plants and animals" or "plants, animals, and microorganisms," thus positing his theories of acclimatization and acquired characteristics as universal.[25] These generalizations were repeated in rural newspapers and circulars, and created the foundation for a successful push toward the modernization of management in animal agriculture. In the case of animals, the popular use of Lysenkoism as a management ideology resulted in small but steady improvements to the scale and productivity of farm animals by influencing labor organization and by essentially micromanaging the task of caring for animals individually.

Most critiques of Lysenko have focused on the disastrous effects his theories had on grain production across the Soviet Union, especially his failure to increase the yield of dryland wheat, a crop that experienced major production setbacks in the 1950s. Historians have typically overstated these failures, claiming that the wildly variable wheat harvests of the postwar period represent a total failure of food security and agricultural modernization. Lysenko's policies certainly contributed to erosion and crop failures across the eastern Soviet Union during the late 1950s. However, the Soviet wheat crop failed numerous times both before and after Lysenko's era due to droughts, heat waves, and political crises, and the uncertainty of grain yields was a source of chronic frustration for the regime. While Lysenkoist policies were devastating to both economy and environment during the 1940s and 1950s, the variable wheat harvest was not a phenomenon Lysenko either created or resolved during his years in power.[26]

Lysenko's attack on Gregor Mendel has also been widely criticized. Historians have correctly noted that Lysenko's critiques of Mendel were politically motivated and that they were unfair distortions of his research. Lysenko's anti-Mendelian stance appealed to his superiors in the Cold War atmosphere of the late 1940s and early 1950s. Regardless of intellectual merit, his critiques pleased Soviet authorities because they implicitly and explicitly criticized prevailing American scientific practices. The Soviet Union first officially endorsed Lysenkoist theory during the summer of 1948, when relations between the United States and the Soviet Union were at a low point. As Medvedev notes in his biography of Lysenko, his downfall

in the Soviet Union coincided with a period of greater international exchange and openness in the natural sciences that
provided a kind of closure to the insular attitude that prevailed
during the height of the Cold War.[27]

Today's society has so thoroughly embraced genetics that
it is difficult to imagine that educated and observant citizens
could have been convinced by a theory of acquired characteristics. However, sixty years ago genetics was a new and un-
proven science, and Lysenkoist inheritance schemes appealed
to common sense rather than scientific knowledge. If, as was
commonly believed, Soviet babies could develop a tolerance
for cold weather by spending a few minutes each day naked in
a cold room, and if warming foods such as vodka and pepper
could counteract the fever and chills symptoms of the common cold, then surely maize plants and young calves might
somehow also learn how to tolerate a frosty Siberian spring.[28]

Lysenko's contrarian system was one false theory among
many that cropped up in the postwar era. During this same
period, Soviet scientists published articles on how to determine the sex of day-old ducks by examining the glands in their
necks and spent millions of rubles researching ESP and teleki-
nesis. These endeavors were certainly examples of bad science,
but they were not particularly closely related to one another, to
the oppressive political system of the time, or to Lysenkoism.
Instead, they were part of a postwar movement to separate
and define Soviet science apart from the scientific culture of
Western and capitalist countries. In other words, these bizarre
scientific projects were artifacts of the Cold War.[29]

The animals bred and raised during Lysenko's tenure at
the Lenin Agricultural Academy were produced with a back-
to-basics set of techniques in which tested, reliable breeding
and handling practices acquired new prestige and were fol-

lowed with rigor by highly supervised workers. Although still relying on techniques developed in the nineteenth century, postwar animal agriculture was not all retrograde; new machines, crops, and animal breeds also arrived in the countryside during this period. However, the Ministry of Agriculture realistically concluded that the mechanization of animal agriculture was a distant goal and should not be a major element of the Five Year Plans that formed the foundational architecture for agricultural policy in the Soviet Union between 1946 and 1955. Scientific management, human capital, and basic, low-tech procedures substituted for machines during this postwar modernization drive. The popular and state-sanctioned policies of Lysenkoism helped this agenda flourish in the postwar countryside. The first sector where Lysenkoism had an immediate impact was in increasing animal populations, the second was in acclimatization projects.

A Lysenkoist paradigm revalued Soviet agricultural assets with a new rubric in which the value of what Soviet animal farms had on hand—elements like human laborers, environmentally well-adapted animals, and effective mass culture—increased dramatically. Other elements that were absent, such as genetic diversity in animal populations and advanced machinery, became less valuable. To rebuild animal stocks the Soviet government became aggressively pronatalist, relying on intensive rather than extensive animal management. The state established and expanded state breeding farms (*gosplemsovkhozes*). These farms created stronger and more prolific breeds and hybrids, often using artificial insemination techniques to speed up and improve fertilization rates, and obsessively tracked the survival rate of offspring in an attempt to increase numbers of animals.[30]

In human populations, pronatalism describes the role nations play in encouraging procreation. Numerous twentieth-century governments, the Soviet Union among them, adopted aggressively pronatalist policies to meet "the demands of industrial labor and mass warfare."[31] The pronatalist policies directed toward animals that the Soviet Ministry of Agriculture embraced after the Second World War had much in common with human pronatalist policies. Increasing the number of farm animals contributed to the health and wealth of Soviet citizens both indirectly (through a more varied and higher-protein diet, and through the increased availability of animal by-products) and directly (through privately owned animals kept for work and food).

In 1948 the Ukrainian minister of agriculture, V. Matskevich, listed the twin problems of livestock fertility and offspring survival as the primary elements that hindered the recovery of cow, sheep, goat, and pig populations in Ukraine. The breeding farms directly addressed these obstacles for Ukraine and the rest of the Soviet Union. Animals produced on state breeding farms were visible, successful results of state-managed reproduction and thus a living form of proof of state achievement and competency. A 1949 annual report from one research and breeding farm listed five ongoing projects: "(1) to create new and improved breeds of animals, (2) to speed up the general rate of reproduction, (3) to increase the productivity of agricultural animals, (4) to improve the nutritional value of feed, and (5) to improve the productive strength of the collective farm system."[32]

Creating new and improved animal breeds was the centerpiece of the work of the state breeding farms. The state focused on creating not just more animals, but also new animals that could fit into specific ecological niches across the Soviet Union. To this end, the breeding farms strived to create some animals that were robustly heat-tolerant, and others that could thrive in extremely cold environments. This project to match animal characteristics to the environmental conditions they might face derived some of its legitimacy from Lysenkoism. From the standpoint of a Lysenkoist, animal breeds were capable of improving permanently over a single generation, thus the work of creating ecologically specific breeds and hybrids was expected to be relatively straightforward.

Soviet breeding and acclimatization programs of the late 1940s and early 1950s show how Lysenkoism became a practical management tool for animal agriculture. Animal breeding stations were locations where ideology and practice merged. They serve as excellent examples of how state desires translated into material realities. In the *gosplemsovkhozes* a uniquely Soviet way of improving animal agriculture took hold where state ideology, socialist work patterns, and environmental and demographic limitations and particularities of the country all contributed to the rise of a new and unique industrial order.

The Soviet Union's most famous animal breeding station, Askaniia Nova, was located on the steppes of southern Ukraine. At Askaniia Nova, Lysenkoist ideology merged with long-standing research practices, and resulted in new animals and new ways of caring for those animals. The Askaniia Nova breeding station became the postwar national centerpiece of the breeding and acclimatization program, and a showcase for how Lysenkoism could be used to improve animal agriculture in the Soviet Union.

Askaniia Nova was originally a massive sheep ranch founded in 1828 by Saxon sheep breeders, who settled in Southern Ukraine at the invitation of Tsar Nicholas I. The fifty-two-thousand-hectare farm became a nature preserve in 1875, although the property's owner, Friedrich Falz-Fein, was forced to sell off most of the pastureland to repay debts. At the turn of the century, Askaniia Nova encompassed eleven thousand hectares and was managed by Falz-Fein's son, also named Friedrich. Although the senior Falz-Fein had originally tried to keep the property as a sheep ranch, his son had success running the farm as a mixed-use enterprise. In the years leading up to the First World War Askaniia Nova supplied horses to the Russian imperial army and housed the finest flock of merino sheep in the empire. The experimental research station Falz-Fein created and managed was also successful. Led by Falz-Fein, workers at Askaniia Nova investigated animal acclimatization, performed horticultural experiments on "virgin" steppes, and created Russia's first wildlife park.

Like many nineteenth-century naturalists and scientists, Friedrich Falz-Fein believed in acclimatization as well as selective breeding, and his park was famous for its collection of Przewalski horses, African gazelles, and horse-zebra hybrids. Falz-Fein was a hobby farmer and something of a dreamer; he owned a wide range of unusual stock because he planned to reintroduce some extinct native species to the Eurasian steppes and also to establish new species in Ukraine. In addition to gazelles, zebras, and wild horses, he focused on building up stocks of flamingoes, gnus, elands, llamas, buffalo, bison, camels, and guanacos. To Falz-Fein, the grassy Ukrainian steppes resembled the savannahs of sub-Saharan Africa, and he imported exotic ungulates and tropical birds, hoping that they would adapt to a new but analogous ecosystem and establish new populations in Ukraine.[33]

Many of the exotic animals Falz-Fein imported thrived at Askaniia Nova, but this was largely because of humans accommodating these new arrivals, rather than the result of rapid adaptation on the part of the animals. Falz-Fein and his staff modified the environment of Askaniia Nova throughout the late nineteenth century. For example, they installed a vast system of irrigation in order to stretch out the growing season and to strengthen the root systems of grasses in the park. Grazing animals were provisioned with hay in the winter and artificial windbreaks sheltered animals during harsh winter storms. The landscape was altered to accommodate new African and Asian animals, and many workers at Askaniia Nova cared for the imported species, ensuring that rare and valued animals would flourish. At its peak of operation in the nineteenth century, Falz-Fein's animal estate employed hundreds of workers engaged in research, grounds maintenance, and animal care. The estate was famous throughout the Russian empire and Europe for its acclimatization, hybridization, and irrigation projects.

The socialist revolution in 1917 altered both the research agenda and the ownership of the park. The new Soviet state seized Askaniia Nova's land, and although it remained protected and was not redistributed to farmers, its scientific research projects were interrupted. Stalinist science priorities effectively ended some of the world's first long-term ecological studies of the cyclical productivity of grassland steppe ecosystems. The state removed ecologists from Askaniia Nova because they argued for sensible measures that would conserve land and water resources. Their arguments and recommendations were based in the (now) familiar ecological theory of natural limits, which contradicted Stalinist notions about the need to exploit nature. Ironically, the setting of this ecological research was just as culturally manufactured, if not as environmentally destructive, as the industrial megaprojects against which this first generation of Soviet conservation scientists argued. Askaniia Nova did not become a site of inten-

sive industrial agriculture until the post–Second World War
period, but the scientifically managed and re-created steppes
that Falz-Fein populated with exotic animals and economically desirable hybrids were a particular kind of colonized and
well-managed landscape.[34]

By the late 1930s there was no ecological research at Askaniia Nova. Scientists now focused on economically relevant
projects in animal and plant breeding. The station changed its
name in step with the times and became the "Trofim Lysenko
Station of Animal and Plant Hybridization and Acclimatization." This administrative change reflected an early triumph
of Lysenko's research priorities, but the station retained its
long-standing reputation for performing solid, well-funded
research on a variety of rangeland animals. The disconnect
that Douglas Weiner and other historians have identified between the work of the research station in the 1920s and early
1930s and the work of the station after 1937 is an artificial
break; Askaniia Nova had a long and well-respected history in
animal breeding and hybridization that predated its studies of
grassland ecology. The Trofim Lysenko Breeding Station could
capitalize on its strength as a landscape that had been modified through irrigation and selective plantings to suit the needs
of a variety of rangeland animals—horses, cows, and sheep, as
well as the more exotic gazelles and zebras. Thus it was in a
well-established setting of accommodation and adaptation that
the Soviet Union's preeminent postwar animal breeding program emerged.

Lysenko's field research and Lysenkoism both conflated
and misused precise scientific terms; a prime example is the
use and abuse of the word "hybrid" at Askaniia Nova in the
postwar era.[35] In the nineteenth century, Askaniia Nova had
specialized in rare interspecific hybrids; that is, crosses that took

place between animals of different species. The zebra–wild horse crosses and crosses between various species (and subspecies) of antelope were famous examples of this work. Fifty years later, in Lysenko's time, other crosses, including grafts and intraspecific hybrids, had become the "hybrids" to which scientific publications referred.[36] This confusion was not unique to Lysenko or to the twentieth-century Soviet Union. Gregor Mendel also failed to recognize the difference between interspecific and intraspecific hybridity in his famous experimental pea crosses, but the mix-up had unusually important consequences in the field of animal breeding. The permissive definition of hybrid that Lysenkoism and Soviet publications supported gave rise to a new series of practices for reestablishing and improving animal breeds. Likewise, early successes "acclimatizing" animals —another vague term—to colder climates provided new techniques for successfully expanding livestock populations into previously marginal environments.[37] The vague terminology allowed the words to represent a broader range of phenomena, and vague words also implied that humans had a high level of control and intentionality over processes of reproduction and adaptation in animals when this was not the case.

This genuine confusion over the meaning of hybridity was both convenient and useful for the Ministry of Agriculture. Hybrid interspecific crosses of the imperial past (such as the zebra-horse) were conflated with the state-ordered intraspecific crosses of postwar animal recovery efforts (such as the Simmental-Polish Red cow). The Lysenko Breeding Station's expertise in the first category meant that the state assumed it would excel in the second category as well. Because of this the station was placed in charge of supervising the reestablishment of several animal breeds, including the Ukrainian White Steppe pig, whose fate is worth examining in greater detail.

In the case of the Ukrainian White Steppe pig, the nine-teenth-century techniques Friedrich Falz-Fein's staff had used to create interspecific crosses and the twentieth-century skill of scientifically reengineering a declining breed of swine were discrete. While the first required an investment of patience, money, and skilled assistance, the second required the biological capital of a sufficient animal population, accurate record keeping, and an authority that could identify and uphold breed standards. In other countries at earlier times, the process of acquiring these professional skills had been time-consuming and energy-intensive, and as a result wealthy private entrepreneurs rather than the government had often initially funded the business of stockbreeding.[38] This was not the case in the Soviet Union, where animal breeding was a state-sponsored endeavor that demonstrated the scientific and managerial superiority of breeding farms and stations across the Soviet Union. For an animal like the Ukrainian White Steppe pig, the postwar approach to improving stock was founded on new Lysenkoist notions of hybrids and acclimatization. This allowed the Soviet Union to obtain what it judged to be animals of superior quality and productivity more rapidly and with less capital investment than was possible under a capitalist program of animal improvement. Unlike the closed and invisible world of genetics research, Lysenkoism was an open science, ideally suited for simple, triumphal displays of progress that put Soviet animal agriculture on display in everyday forums like newspapers and posters.

Organized work on breeding the Ukrainian White Steppe pig peaked in the late 1940s and then fell off sharply in the early 1950s, when authorities redirected their efforts toward creating a new "speckled" breed rather than resurrecting the older, traditional white breed, because the speckled pigs gained weight faster and Ukrainian White Steppe pigs frequently experienced bronchial infections.[39] Although the Ukrainian White Steppe pig ultimately became an evolutionary dead end, efforts at improving the breed were typical of the era. This case study shows how Lysenkoist ideology pervaded state-sponsored animal husbandry and supported an agricultural policy that was both pronatalist and dependent on the skilled labor of women.

The Ukrainian White Steppe breed was first created before the Second World War as a dual-purpose meat and lard pig. The war devastated all porcine populations, but the Ukrainian White Steppe pig fared especially poorly; in Ukraine the postwar population was approximately 30 percent of what the prewar population had been.[40] In planning the recovery of the breed, livestock managers at Askaniia Nova anticipated three major problems. First, there was the organizational difficulty of coordinating a breeding program, as the majority of the pigs were located not on Askaniia Nova proper, but at several farms in the vicinity of the station. The second challenge was the very basic task of improving the daily food supply of the animals, and the third issue was improving the breed through better selection.

That the Ukrainian White Steppe pig was chosen for recovery at all was due to Lysenkoist thinking at the highest levels of the Ministry of Agriculture. Lysenkoism posited that each agricultural region of the Soviet Union needed animals adapted to its unique ecology. Breeding stations sometimes developed animal varieties for a single department or even a single district. The resurrection of the Ukrainian White Steppe pig was a part of this movement toward ecological specificity. The breed was native to the southern districts of Ukraine, and according to official plans, it was supposed to repopulate two districts in Kherson Oblast in eastern Ukraine. However, as a hastily and recently developed breed, the Ukrainian White Steppe pig possessed several endemic and undesirable traits including weak leg joints and a tendency to give birth to feeble piglets. In hindsight, this was not an ideal breed to single out for vigorous restoration efforts.[41]

The project to restore animal populations that had been lost in the war parallels the state's shifting attitude toward postwar reconstruction in Ukraine regarding humans. Many Soviet families chose to start over and create new communities in other parts of the Soviet Union after the war, but throughout the late 1940s state policies still focused on re-creating population centers that had existed in Ukraine before the war. It was not until the 1950s that the Ministry of Agriculture accepted the demographic realities of postwar Ukraine and shifted efforts away from rebuilding toward new industries and new settlements further east.

Coordinating a breeding program that occurred simultaneously on several farms was challenging for the Ministry of Agriculture. The Lysenko Breeding Station reported in 1947 that it had "strengthened the methodical management of selection work" to solve logistical issues it had encountered the previous year.[42] This abstract phrase, "methodical management," conceals a real work process that is interesting to understand. What changed in the work of the station in 1947 was first, that record keeping improved, and second, that the station increased the number of workers and the overall skill level of workers hired. In essence, "methodical management" meant having more people do more work, and recording this work in an organized fashion. The report then describes the projects of the most skilled researchers and technicians at the institute, grouped by project theme under headings such as "pigs," "horned livestock," "chickens," and "feed." Financial reports from the Lysenko Breeding Station in 1947, 1953, and 1956 also establish that a significant and increasing amount of money was paid out to twenty-two supporting collective farms in the area for services rendered. These supporting farms provided laborers who would work with the animals at the station, and in some cases, the animals were housed on neighboring collective farms; Askaniia Nova paid these farms a boarding fee for housing and caring for pigs and other animals. It is a reasonable guess then that, as was common in the postwar period, Askaniia Nova was indirectly supporting the salaries and in-kind payments of newly valued Soviet farm workers called *svinarki,* or female swineherds. In the postwar era, *svinarki* were instrumental in improving animal survival rates on smaller farms as well as at major operations like the Lysenko Breeding Station.

The position of female pig tender developed because of the skewed demography of the Soviet countryside in the wake of the Second World War. In many areas, women outnumbered men by a ratio of two to one, because of the high casualty rate of the Soviet Red Army and the slow rate at which soldiers were demobilized at war's end. During the war women had taken over men's positions on the farm, such as plowing, harvesting, and running grain collection facilities. Women remained on the payrolls of kolkhozes and sovkhozes in a variety of traditionally masculine jobs, but beginning in 1948, as military demobilizations increased, men returned to these better-paid positions and women workers were in surplus. Animal care, which had never before been a significant labor category on farms, became an increasingly important sector of rural work into which collective farms could siphon this surplus of women.

Women had always worked on animal farms, but after the war their numbers soared. Both the Fourth Five Year Plan, adopted in 1946, and the revised Three Year Plan for Agriculture of 1947 called for intensifying and improving the level of care and feed that collectively held animals received, and to do this these farms created more positions for animal caretakers like the *svinarki*. Lysenko addressed this new policy focus directly in an often-repeated statement that "the basis for increasing the productivity of domestic animals, for improving existing breeds and producing new ones, is their food and the conditions in which they are kept." *Svinarki* and women who worked with other animals became the direct overseers of these food and material conditions.[43]

The theory of single-generation breed stabilization and rapid productivity increase was based on Lysenkoism, but when this theory succeeded it was because of the work performed by women. *Svinarki* improved piglet survival rates first by adhering to nationally prescribed rituals of daily care. These instructed workers in the most basic kinds of care, including checking on sick animals every day and keeping their sheds and stalls clean. The national standards also included more modern specifics, such as adding fish oil to the feed of the weakest animals and heating the rooms where malnourished litters slept.[44]

The second contribution *svinarki* made was in record keeping. Part of their job was to chronicle the progress of their charges by making regular log entries for individual attributes such as weight, temperament, and feed consumption. As a corollary to this project, pigs needed to be positively identifiable as individuals. This gave rise to a standardized system of naming and numbering animals. Pigs were usually identified by parentage, litter order, and the year in which they were born, by a series of notches or tattooed marks in their ears.[45]

By quantifying the experiences of their charges and itemizing the animals themselves, *svinarki* provided a window into an otherwise opaque world for outside office-level authorities, such as the authors of Askaniia Nova's annual report, who were able to browse through such records and translate the lives of pigs from twenty-two separate farms into unified narratives of general progress and modernization for the benefit of their superiors in Moscow. This same project of quantification allowed bureaucrats to decide which local pigs could be considered foundational stock. For pigs of uncertain parentage, farm managers created lists comparing the size, weight, coloring, and disposition of animals, and authorities then decided which pigs among these exhibited the true characteristics of the breed. The word was not always law, however, and in at least one case, a committee of experts made an on-site visit to a farm in order to observe the Ukrainian White Steppe candidate. Indeed, although historical perspective shows that the Ukrainian White Steppe breeding program ran into trouble within just a few years because of inherited breed weaknesses, until the program was canceled in 1959 there were few indications from the research station that the program was experiencing anything aside from phenomenal success.[46]

Central authorities also ruled on the amount of time pigs should spend outdoors (four to six hours in summer and two outings of twenty minutes each in winter), and recommended constructing smaller dens rather than spacious stalls for pregnant sows so that they would feel more comfortable indoors.[47] These standards of practice were disseminated in rural places through preexisting media, including educational films, mandatory workshops with vet-feldshers, night classes, and illustrated posters; local Party committees coordinated their distribution.

Perhaps the most detailed instructions came from articles in local, Party-run newspapers that repeated stories of the daily lives of successful *svinarki,* focusing on their dedication, sobriety, and civic-mindedness. These stories highlighted the maternal care these women gave to their charges and their pride in their job in spite of challenges.[48] The message of work expectations was clear and constant: pig caretakers were personally responsible for the fecundity of their sows and the survival rate of litters. While incentives such as free piglets were occasionally given out as rewards for extraordinary labor, low productivity or high mortality rates were a sign of poor performance. Contrived socialist competitions between *svinarki,* in weight gain or litter survival rate, were given more column space, but the real proof of the job's expectations were found in the "shaming" sections of the papers, where drunken, slothful, or otherwise inept animal caretakers were chastised publicly, especially when their poor job performance resulted in the death or illness of an animal. The continued seasonality of pig management into the 1950s is also obvious from newspaper articles: *svinarki* biographies appear most frequently in the late winter and early spring when pig mortality was at its highest. The stories disappear in the late spring and summer months, replaced by more seasonally appropriate articles on productivity in field labor and the arrival of new farm machines.

One obvious way to note the prevalence of Lysenkoist ideology in the commonsense and labor-intensive measures bureaucrats required from *svinarki* and other animal caretakers were how these requirements paralleled practices in plant agriculture. Although plant and animal agriculture were institutionally separate organizations, treatments ordered for improving seedlings and baby animals were similar. The vernalization of maize, the "airing" of pigs in the winter, and the "cold method" of calf raising all advocated the same approach to helping three different species overcome their natural disinclination to cold weather. The practice of environmentally tailoring seeds and animals to specific microclimates around the Soviet Union was also identical; seed varieties and animal breeds were designed to succeed in a very narrow geographical band. In a reversal from practices elsewhere in the world, seeds and animals were often developed at research stations thousands of miles away from their intended new homes.

The work of *svinarki* made a difference both socially and materially. These women combined existing local knowledge of how to care for pigs and their offspring with new and more bureaucratically influenced instructions from their superiors. Nursing runty piglets, warding off winter bronchial infections, force-feeding young stragglers, and hosing down pens made a significant difference in reversing previously abysmal survival rates, thus increasing the established success of postwar pig operations. Likewise, the seemingly banal work of taking stock of pigs by name, genealogy, growth, temperament, and appearance had important ramifications above and beyond the imaginary world of Lysenkoist breed standards and regional annual reports. An itemized pig was potentially a healthier and more productive pig. While the systems of indicating identity such as ear notching and tattooing could initially be physically traumatic (deafness was common), barnyard legibility had its benefits. Written records created medical and family histories that helped keep track of increasingly complicated feeding regimes, growth patterns, and breeding cycles. Marking pigs with names and numbers allowed their caretakers to care for their individual life situations, even if the caretakers had no personal experience with the animals in question.

While the care and improvement of the Ukrainian White Steppe pig offer certain insights into the relation between agricultural management and Lysenkoism, the history of postwar dairy cows illustrates the ways in which Lysenkoist assumptions about breeding and inheritance increased the value of

Figure 1. A *svinarka* at work. Used with permission of Iowa State
University Library, Special Collections Department.

Soviet livestock. In the area of pig improvement, Lysenkoism
revalued groups of people, such as *svinarki*, whereas the history of milk cow improvement is a story about the ways in
which Lysenkoism revalued certain animals. Simply put, the
confused Lysenkoist interpretation of hybridity allowed large
numbers of dairy cows of uncertain parentage to attain purebred and hybrid status in a short period of time. Generally
speaking, and certainly in the case of the Soviet Union, hybrids and purebreds were worth more than mixed-breed animals (or to use the Soviet terminology, *metis*), thus this evolution of pedigree meant an increase in livestock value, but more
fundamentally, a change in the way livestock were valued. Just
as in the case of the female swineherds, milkmaids or *doiarki*
played a central role in improving the productivity of cows,

especially in the realm of milk production. The path from *metis* to breed was often configured along lines of just such productivity, thus old jobs acquired new status and new work practices resulted in newly valued animal categories.

The progression from *metis* to established breed was common enough in the postwar era, but there was no prescribed series of steps from one category to the next. Promotion was sometimes decided by physical appearance, sometimes by a partially reassembled bloodline, and often by performance. The decision to promote was made by experts from outside the host farm or breeding station. It is unlikely that the work of "improving" breeds of milk cows, invented in 1945 and promoted heavily between 1945 and 1948, intended for Soviet animals to be proclaimed purebreds and scientifically (that is, intentionally) produced hybrids as rapidly as they were in the 1948–55 period.[49] In 1945, Lysenko was a peripheral figure in animal science, famous for his experiments with wheat, but not yet established as a leading voice of authority in matters relating to scientific breeding. This had changed by 1948, when the buzzword "hybrid" entered the consciousness of the middle-level bureaucrats who oversaw breed improvement programs. The best way to examine the path through which a mongrel milk cow might attain *porodnost'*, a Russian word best translated as "purebred status," is to examine some real-life examples.

The first example comes from Askaniia Nova, which was charged in 1945 with improving Simmental cattle, a milking breed originally developed in Switzerland and imported to Ukraine and Russia (along with potatoes) by German immigrants during the reign of Catherine the Great. However, based on newspaper photographs and production statistics, the Soviet Simmental of the mid-twentieth century diverged

sharply from its Swiss ancestors in both appearance and productivity. (Today, Simmental breeds in Russia and Ukraine
still average only about two-thirds the size of their Western
European counterparts.) The breed resembled the cattle described by Americans at the beginning of the chapter: a small,
furry, and nondescript multipurpose animal. The Simmental
were the most numerous among the minority of Ukrainian
cattle that were classed as a breed at all (the vast majority, in
1945, were listed as not having purebred status); they were
more commonly referred to as *shvitskoi* or *metis-shvitskoi*—
"Swiss-like" or "mixed, Swiss-like." At Askaniia Nova the two
actions taken to improve the Simmental line were breeding
work and improving the diet, in order to breed a "new type
of high-producing animal."[50] That these two tasks might be
considered equivalent measures for breed improvement is surprising on its own. More surprising are the changes in policy
reflected by reports from the station published after 1948 (that
is, after Trofim Lysenko's rise to power).

After 1948, one branch of the work to improve the Simmental cow evolved into a push to create an entirely new breed
of milk cow, the Lebedinskii breed. The Lebedinskii Raion was
a short-lived administrative region located in the present-day
Belgorod region, directly north of the city of Kharkov. Because
it had a lot of grassland, Lebedenskii Raion contained many
Swiss-like and *metis* cattle, many with Simmental ancestors.
Beginning in 1948, the Ukrainian Department of Animal Agriculture decided that these animals were good candidates for
promotion to the status of purebred Lebedinskii cattle. Before
1948, the Lebedenskii cow did not exist; indeed, before 1945,
neither had the Lebedenskii geographical region.[51]

In the case of the Lebedinskii breed, purebred status
was not based on stud books or even on phenotypical appear-

ance. Instead the breed was created through paperwork and a site visit by agricultural authorities. In 1949, one major action for "improving the cattle of the Lebedinskii breed" was the "preparation of materials toward establishing the Lebedinskii group as an independent breed and a statement of a verifying inspecting body." The "inspecting body" traveled from Kharkov to Lebedinskii Raion and observed thirty-eight hundred cattle, deciding unanimously that all them should be counted as a new breed. Establishing a new breed meant not just giving cattle that had been considered Simmentals or mixed breeds a new, geographically oriented, and Slavic name (although this was not insignificant). It also established breed improvement as a bureaucratic, rather than a field activity. Although the "inspecting body" had traveled to Lebedinskii in order to ascertain the *porodnost'* of the cows living there previously identified as mixed-breed, their unanimous decision to establish the breed implies a foregone conclusion; witnessing the cattle in situ was a formality necessary only to conform to expectations of scientific objectivity.[52]

Some breeds were of greater interest than others for the Ukrainian Ministry of Agriculture, largely due to their levels of productivity. For example, the Simmental and Ukrainian Grey Cattle (another breed selected for a radical improvement regimen in the postwar period), were known to be more generally productive than the scrubby animals American visitors spotted from their train windows. In general, increased productivity (measured in liters of milk per year or by butterfat percentages) was linked to animal breed status. Award-winning animals were never listed as "breedless" when they were displayed at agricultural fairs. Between 1945 and 1947 they were listed as *metis*, and after 1948, as hybrid animals or new breeds. While Lysenkoism erased some of the historical value and power of elite genealogy, it did not remove it completely. Instead, it substituted a language of scientific breeding that implied that the bloodlines and productivity levels of animals were under the purview and careful management of the state. Collective farm chairmen and Ministry of Agriculture scientists often presented themselves as the managers of these new and improved cows, but this was false advertising. In fact, weight gain in calves, milk production, and butterfat levels were variables over which another kind of manager—the Soviet milkmaid—had control.

Just as with pigs, there were national standards of animal care for dairy cows that Soviet *doiarki* were expected to follow, and which were designed to maximize milk output from each cow. Many of these standards prescribed specialized and rigorous hand labor, which was intended to make up for the lack of machines in Soviet milking parlors. Cows received an udder wash and a three-to-five-minute udder massage to stimulate milk production. In the winter, goose or bear grease was applied to the teats after every milking session so they would not freeze. The milk that had been collected from each cow was weighed, and if possible, the butterfat content was tested. The weight and fat content of the milk were recorded in a registry book, and the next cow came up for milking. Some collective farms had milking parlors and stalls dedicated to this work, but it was more common for milkmaids to do their milking in a quiet corner or stall of a common barn. This practice was one of the first things to change in the period of machine industrialization that followed, but between 1946 and 1955 the Soviet Union did not typically have the building resources to provide every small dairy with an additional building for milking. In the winter, cows were supposed to be taken on a *progulka,* or stroll, outside to stimulate milk production. After the morning walks milkmaids tended to sick animals and weighed the calves before the midday milking, when the whole cycle began over again, and was repeated one last time in the evening. Milkmaids typically worked twelve-to-sixteen-hour days. Additionally, when a cow fell sick, milkmaids were expected to come in early to provide extra care, or to sleep over in the barn's sick ward. These kinds of prescribed work routines provided a framework that allowed greater human intervention in the process of breed improvement as well as in that of stock increase in the immediate postwar period.

Proper sanitation, heated water, and periodic exercise for dairy animals were not innovations in the late 1940s; they were simply given a new emphasis. In its heavy borrowing from the past, postwar animal agricultural policy repeated a tactic used many times by the Bolsheviks. In spite of this back-to-basics approach, postwar agriculture was not a particularly retrograde system because its successes lay in the skill with which it combined old-fashioned ideas with recent advances and above all with a language of precise management and continual advancement. Just as in the case of animal breeding, agricultural policies for animal care capitalized on the Soviet Union's agricultural strengths. Short on machines, the Soviet countryside was long on semiskilled female laborers. Likewise, the Soviet Union lacked well-pedigreed farm animals, which were the standard measure of the capital worth of livestock in countries that had industrialized their agricultural systems. However, armed with Lysenkoist ideology, the Soviet state believed it understood how best to manage and breed the animals it did possess toward maximal productivity. It correspondingly set about redefining the qualities that made livestock valuable at the same time as it emphasized the human dimensions of animal care.

Historians have long wondered how a system such as Lysenkoism, which did such a poor job describing and predicting the life experiences of plants and animals, could have survived for a quarter-century as the dominant biological paradigm in a modern state committed to an ideology of rational, efficient production, especially a state in desperate need of a better and more productive agricultural system. The standard explanation of this confusing state of affairs has been to condemn the corrupt and incompetent political structure of the Soviet Union that placed politics over accuracy, and ideology over experience. A better explanation for the long life of a bad science is that Lysenkoism made bad experimental science but good practical science. It did not provide an accurate model of what happened to plants and animals as they bred, lived, adapted, and died, but it provided an excellent set of instructions for what semiskilled workers could do to assist these life processes with limited resources.

Lysenkoism described appropriate ways in which workers and scientists could engage with—and improve—the natural world.[53] These actions were not just ideologically suitable to the narrow Marxist-Leninist rhetoric of the day; they were also in line with the state's desire to overcome historic environmental limitations on animal agriculture. The long-standing evolutionary adaptations domestic animals had made to cold weather and scarce feed were traits the Ministry of Agriculture believed Lysenkoism could undo in less than twenty years. Lysenkoist management techniques transformed Soviet farm animals into creatures that had, by Soviet standards, the right size, shape, and temperament to boost meat and milk productivity so that the country could compete with the much more technically advanced system of the United States.

Historians and social scientists typically describe the So-

viet Union's postwar recovery as slow and relatively ineffective, hampered by poor organization and the damaging rise
of Lysenko's backward theories. This was certainly true for the
first two years after the war, which saw extreme food privation
and famine in some areas of the country. By late 1947, however, the country entered a period of stabilization, recovery,
and growth. Measured against the expanding economy of the
United States, the Soviet Union's postwar renewal does not
look impressive. Yet rapid expansion is not always the best
kind, and while the Soviet Union did not often meet its ambitious goals in the realm of animal agriculture, the gains the
country made in the five years after the Second World War
were impressive.

In studying agricultural efficiency and productivity, the
ability to produce staple carbohydrates has long been the gold
standard for predicting success. The Soviet Union's inability
to increase the production of grain is often mentioned as an
Achilles' heel of postwar agriculture. But beyond grain, Soviet
agricultural policy and state production campaigns focused
an unusually large amount of their energy and finances on
animals. Although the theory behind any supposedly rational
modern diet assumes that high-input animal products such
as meat and milk are luxuries in developing economies, the
Soviet Union's attitude toward animal products was not always
rational, and thus milk and meat production occupied an unusually important and symbolic place in the hierarchy of national agriculture.

The story of Soviet meat and milk production also offers
a contrast with the standard story of agricultural intensification and modernization. Humans, not machines, were instrumental in the expansion of animal agriculture. Postwar animal
breeding and caretaking policies provided a system through

which the Soviet Union was able to temporarily surmount the environmental and technological barriers that had prevented the state from industrializing animal agriculture in the previous decades. Lysenkoism provided a temporary foothold for the state, an irrational but functional system of management that effectively matched the skills and strengths of the rural human population with the needs and challenges of the Soviet Union's farm animals and environment.

4

Substituting Meat

The comestible afterlives of the raw agricultural products of milk and pork also represented a problem and an opportunity for the Soviet state. Just like the new farms that focused on raising more animals in the countryside, food factories in the cities were enjoying a renaissance, expansion, and reorganization during the 1950s. The rebirth of meat and milk in a Soviet context is best illustrated by two new prepared foods that appeared in stores during this decade. The first, *tushonka*, is a processed pork product that distantly resembles Spam. *Tushonka* became a ubiquitous consumable for a number of years after the Second World War when Soviet planners realized that processed foods with a long shelf life could help them circumvent issues of timing and orderly distribution. In the process of creating more *tushonka*, the meat industry shored up the new factory-style farms that Lysenko's scientific influence had created, ensuring that in spite of their inefficiencies, these new facilities would hold a permanent place in Soviet food production.

A second example, ice cream, demonstrates another creative solution to the problem of food timeliness. With *tushonka*, Soviet planners simply evaded the question of spoilage by making a food that could (almost) never decay. With ice cream, planners were forced to change the system rather than the product, and the result was a stunningly successful network of pushcart ice cream vendors that citizens of the former Soviet Union still remember fondly. Soviet citizens did not live exclusively on ice cream and canned meat, and several other examples of processed foods—some of them animal based, some of them not—are also of relevance here, since they helped ensure the new and improved postwar food system actually worked. Finally, just as farms and fields were restructured to accommodate new forms of socialist production, shops and kitchens also experienced their own minirevolution. In the case of food processing and distribution, the Soviet Union was far more successful with its projects of modernizing and scaling up than it had been in the agricultural sector.

Before 1948, the food sector had been of secondary importance in Soviet national policy because it was a so-called "light" industry, and state propaganda (as well as Stalin himself) preferred to focus on the heroics of heavy industrial endeavors, such as steel and coal production. This attitude changed after the Second World War, when Soviet policy steered away from the working "cafeteria consumer" of the prewar period and toward a recognition and acceptance of a new, more private "kitchen consumer." The leader most closely associated with this newly fashioned kitchen consumer is Nikita Khrushchev. In fact, change started earlier: It was Stalin who first voiced the policies that pushed ice cream, *tushonka*, and other processed foods toward consumers. Stalin, in turn, was nudged in this direction by one of his most trusted advisers, the Minister of Foreign Trade, Anastas Mikoian, a champion of new industrial food processing. It was not just foodstuffs, but also furniture, bicycles, kitchen appliances, and fashionable clothing that light industries produced in this period. All these products were meant to quell potential unrest and restore public faith in socialism.

In the process of introducing such products, Soviet planners also modified their message to citizens about what to expect in the future from socialism; Stalin's often repeated 1936 promise that the future would be "more joyful" still held true, but as of 1946, the future would also be sweeter, saltier, and more comfortable. This vision of anticipatory socialism became one in which conveniences and luxuries helped to prove the success of the socialist model. Where once even the most banal middle-class possessions—a rubber plant, a jazz album—were suspect bourgeois entrapments, in the new postwar era, modest affectations of plenty and middle-class life were tolerated and even promoted by the state. Khrushchev may have mocked the extravagance of American kitchen appliances during the 1959 "Kitchen Debate" at the American Exposition in Moscow ("Don't you have a machine that puts food into the mouth and pushes it down?" he asked Nixon, as he walked past a dishwasher), but by the 1950s, the Soviet Union had largely abandoned its vision of a citizenry dependent on cafeterias and mess halls that had been so popular in the 1930s.[1] Instead, state planners focused on bringing identifiably socialist foods, appliances, recipes, and consumption patterns into the hearts and homes of its citizens.

In the years after the Second World War the Soviet national diet reflected a new relationship with food for the country as a whole, based not on the threat of scarcity, but on the promise of emergent prosperity. In the 1950s, American citizens were well aware of the statistic that the average American ate more meat than the citizen of any other country, but in the

Soviet Union there was no such public statistic. What was obvious to the average Soviet consumer was that meat and milk consumption rates were going up, and that the rations and privation of the immediate postwar period were rapidly disappearing. While consumption levels of certain products remained level or even decreased slightly in the decade after the war, access to animal-based food products increased overall. Increases, real or imagined, in production were trumpeted in the Soviet press, with articles inevitably promising that much greater increases were anticipated in the future. Such reporting had the effect of focusing citizens on what was yet to come.

The postwar preoccupation with producing more meat, milk, and butter, while inspired by a sense of competition with U.S. farm productivity, was part of a larger national project to improve public nutrition, a program of "recovery plus" from wartime scarcity. This movement was hardly unique to the Soviet Union. The United States and the countries of Western Europe all engaged in campaigns to improve standards of living and increase the availability of consumer products. These campaigns were often influenced by Cold War tensions, with both capitalist and socialist countries out to prove that the way of life they promoted was superior.[2] The campaign for more meat and dairy products, although the most publicly advertised, was not the only campaign for better nutrition. Other foods had special resonance in the postwar era, especially those that helped stave off the other scourges of malnutrition. Thus, the Soviet Union's new citrus and canned tomato industries, and its new frozen prepared foods such as raviolis and croquettes, all received a boost from the same kind of postwar dietary preoccupations.

The postwar Soviet Union excelled in its ability to harness and refine the energy sources that defined a developed and high-tech civilization. The hydrogen bomb, first tested in 1955, was the standard par excellence, but other high-tech projects to refine rocket fuel, develop jet propulsion technology, and build hydroelectric dams were pursued energetically and successfully in the postwar period. Of course, canned meat is a more prosaic form of potential energy than rocket fuel or enriched uranium, but the florescence of processed meat alongside its more spectacular cousins in the postwar era is no coincidence—all are historical artifacts of a period that privileged concentrated, quantifiable displays of power and wealth. Whether measured in calories, megatons, or kilowatts, energy was on display in these modern systems and the ability to effectively harness natural energy and make it available to everyday people in order to improve the quality of their lives was one of the more humanitarian fronts upon which the Cold War was fought. Meat and milk production signified wealth and prosperity, but also national strength.

Dislocations of supply lines during the Second World War changed patterns of consumption and the availability of meat and animal products in the Soviet Union, first by reducing supplies and then by forging new lines of distribution through the selective rebuilding and new construction of roads and rail lines. The war profoundly crippled meat and milk supplies, but wartime provisioning measures also increased the use of canned foods, first as a front line provision for the Red Army, and then as a relief and recovery product for the civilian population.[3] In the postwar era, reliance on canned products endured, primarily because they circumvented timing issues that the Soviet Union constantly struggled with in its supply chains. Canned foods did not spoil quickly, and so they could sit in train cars or on grocery shop shelves for weeks or months. This was in marked contrast to the struggles the Soviet Union experienced supplying fresh milk or fresh meat to any but the most centrally located settled areas. Although canned luxury foods such as crabmeat and caviar had existed since the nineteenth century, the postwar era was the first time canned products were made available and affordable to the general population through state stores.[4] While canning vegetables and fruits continued to be a home-based activity for many citizens after the war, canned meat in particular acquired new prominence in the postwar Soviet diet.

Between 1943 and 1955, *tushonka* rose from obscurity to become an emblem of socialist modernity. Demand for *tushonka* shaped production lines, the layout of farms, and even the genetic stocks of the animals that would be eaten. State planners loved *tushonka* from the start because it was inexpensive to manufacture and easy to transport, and it served as tangible evidence that the state was realizing its goal to get more meat to its citizenry. By the mid-1950s, *tushonka* had become a symbol of the Soviet Union's much-vaunted push to establish a modern food regime to rival that of the United States. From its initial position as a Second World War lifeline to its subsequent ubiquity in Soviet markets, *tushonka,* as well as other processed meats, was not simply a product of Cold War posturing, it also influenced how food was produced. Between *tushonka*'s first appearance in 1943 during the war, and the 1957 announcement by Khrushchev that all livestock farms would be structured around the mass production of one or several meat and milk products, *tushonka* had become a universally available product. It was *tushonka*'s success as much as Khrushchev's admiration for American-style pig farms that established the factory farming system in the Soviet Union.

The Soviet Union's first request for canned pork from the Americans came in 1943, during a time of severe food shortages. *Tushonka* was an emergency ration that was meant to sustain soldiers in the field and alleviate malnutrition for civilians. The first cans of *tushonka* were made in the heart of the American Midwest, at meatpacking plants in Iowa and Michigan.[5] Its precursor was its more famous cousin, Spam, which had started appearing on American plates in 1937 and, like *tushonka,* played a heroic role in the war. United States Army contracts supplied Spam to outposts as far-flung as Britain, Hawaii, and Papua New Guinea. For U.S. meatpackers, Spam and *tushonka* were well-timed products; the United States had a surplus of pigs in 1943 and 1944, and shipping pork overseas was a practical way to increase business and raise meat prices by making the product scarcer at home. Unlike Spam, which acquired part of its status from being an innovative product, *tushonka* was a food with a past. The recipe originated in a remote region in the Ural Mountains. Rather than using modern canning techniques, which preserved meat by applying pressure and heat to a sealed jar or can, *tushonka* was originally conserved through the addition of copious amounts of salt and lard. Thus *tushonka* was first conceived of not as a convenience but as a traditional and familiar food; a taste of old-fashioned home cooking soldiers could carry with them into the field.

Tushonka's ubiquity in the postwar Soviet Union initially had little to do with the country's own pork industry and everything to do with taking advantage of American leftovers. Soviet pig populations had been decimated by war, and pigs that survived the Axis invasion were evacuated east with human populations. During and immediately after the war, the Soviet Union was in no position to domestically mass-produce *tushonka*. Instead, *tushonka* was plentiful in the pig-scarce Soviet Union as a result of President Truman's unexpected September 1945 decision to end all "economically useful" Lend-Lease shipments to the Soviet Union. By the end of that month, canned food was one of the only products still being shipped as a Lend-Lease supply to the former U.S. ally.[6] Although the United Nations Relief and Rehabilitation Administration distributed American food supplies to ensure they reached the neediest aid recipients, the Soviet postwar entrepreneurial spirit was strong, and travelers to the Soviet Union in 1946 (and as late as 1950) remarked on seeing cans of *tushonka* for sale in state shops as well as by private individuals.

Parts of the Soviet Union remained awash in donated American *tushonka* for years, but it was only in 1948 that Soviet policies called for domestic production. A crackdown on private trade that began in April of that year complemented increased operating budgets for meat factories and municipal freezers.[7] Because it was shelf-stable, wartime *tushonka* had served as a practical food for soldiers, but after the war, *tushonka* became an ideal food for workers who had neither the time nor the space to prepare a home-cooked meal. The Soviet state invested in *tushonka* because it was such an excellent fit with the needs and limitations of both the Soviet state and Soviet homes. It was a practical way to capitalize on the kind and cut of meat the Soviet Union had in the (relatively) greatest abundance in the 1950s—low-grade pork scraps.

Just like *tushonka*, pork products such as sausages and frozen raviolis also became winning postwar foods, thanks to a happy synergy of a burgeoning pig population, new standards of grading and dismembering meat, and new food-processing machines. As pigs increased in both population and mass, so did the processed meats industry. One official source listed twenty-six different kinds of meat by-products available in 1964, although not all of these were made from pork.[8] Another publication described how meat shops should wrap and display sausages, and listed twenty-four different kinds of sausages that stores might receive.

In the face of a shortage of packaging and labeling, the string that bound the sausage was wrapped in a different style for every type of sausage, and shop assistants were trained to be able to identify sausages based on the patterns of their binding. Consumers, who received no such training, relied on experience, and presumably learned from their mistakes. Raviolis were produced, according to Soviet sources, at every factory that processed pork, using an automated process in which they were "made from start to finish in a special automated machine; human hands do not touch them, which makes them a higher-quality and better product."[9] This aversion to handmade is a theme that reappears with great frequency in the Soviet literature on processed food production.

The Ministry of Agriculture rarely recognized antecedents for its food and farming policies, but the experiences of Soviet pigs strongly resembled those of their British and American counterparts generations earlier. It was in Britain in the 1830s that farmers first recognized the economic advantages of setting pigpens alongside breweries and distilleries, so that pigs could grow fat on the waste products of these industries. American and German farmers soon adopted this system. With pigs now eating a set, grain-based diet, new traits such as speedy growth and robust health were actively sought out in breeding populations by farmers, and later by scientific breeding programs. Pigs had always tended to be sickly and slow-growing, but their ill health and leisurely pace of maturity were balanced by the fact they were easy to keep around; unlike sheep, cows, and horses, pigs did not need pastureland upon which to graze, and unlike poultry, their meat had a high fat content and could be smoked, salted, sugared, or canned to extend its life.

Pig breeders on all continents soon discovered that pig traits had a tendency to split into two phenotypically distinct groups. One of the major breeding debates of the early twentieth century was whether pigs should be bred for "hot blood" (in other words, fast maturation and prolific reproduction) or "big type," a self-explanatory descriptor. Breeds tended to excel in one or the other, but not both, of these traits and it was a matter of opinion whether size or alacrity was the most desirable trait for industrial pigs. Overemphasis on either set of qualities damaged survival rates. At their largest, big type pigs resembled small hippopotamuses, and sows were so corpulent that they unwittingly crushed their piglets. But the trimmer hot-blooded pigs had a similarly lethal relationship with their young. Sows often produced litters of upward of a dozen piglets and stressed mothers often devoured their offspring in a time-tested culling technique. Nonetheless, by midcentury, American pig breeders had all but given up the quest for big type pigs, and focused instead on leaner, faster-growing animals that could get to market quickly.[10]

Belatedly, in the early 1950s, Soviet pig breeding—informed and confused by a flawed Soviet science of inheritance and inbreeding—decided the best approach was to figure out how to maximize both traits, resulting in a decade-long hunt for the elusive "Big Hot" pig. Regardless of the finer details of Lysenkoist breeding practices, well into the 1960s Soviet meatpackers were slaughtering and processing older, fatter pigs with more fat than lean meat on their carcasses. As with many technological items, Soviet pigs lagged behind their capitalist-world counterparts for most of the twentieth century, and Soviet planners in both the food and animal husbandry sectors were keen to catch up. Soviet pigs had been handicapped by cold weather, long winters, poor organization, and poor feed regimens, and they had been valued more for their fat than their meat. Until the mid-1950s, the most commonly bred and most valuable pig on many Soviet state and collective farms was the "lard and bacon" pig. Beginning in the 1950s, the Ministry of Agriculture tried to divorce itself from the this type of pig, with only moderate success. American visitors to Soviet slaughterhouses and farms in 1955 noted that Soviet pigs were far fatter and more old-fashioned–looking than their American counterparts. Nevertheless the 1950s were marked by a dramatic increase in the number of pigs in the Soviet Union, as well as by a much greater diversity of "improved" breeds across the country.[11]

It was not just the status or organization of genetic stock that received a makeover in the postwar drive to can more *tushonka*; a revolution in the barnyard also took place, and in less than ten years, pigs were living in new housing stock and eating new types of feed. The most obvious initial postwar change was in farm layout and the use of building space. In the early 1950s, many collective farms were consolidated. Two, three, or four collective farms were combined into one economic unit, thus scaling up the average size and productivity of each farm and simplifying their administration. While there were originally ambitious plans to recenter farms around urban-patterned population bases with new, modern farm buildings, these projects were largely abandoned as expensive and impractical. Instead, existing buildings were repurposed and the several clusters of farm buildings that had once been the heart of separate villages acquired different uses. The brave new dairy and swine stock developed under Lysenkoist influence did not just have new genes and new regimens; they also had a new architecture focused on their growth and development.[12]

Barns were redesigned and compartmentalized on principles of gender and age segregation—weaned baby pigs in one area, farrowing sows in another—as well as with the aim of maximizing growth and health. Pigs spent less time outside and more time at the trough. Animals destined for different purposes (breeding, meat, and lard) were kept in different areas, isolated from each other to minimize the spread of disease as well as improve the efficiency of production. Much like postwar housing for humans, the new and improved pig barn was a crowded and often chaotic place where the electricity, heat, and water functioned only sporadically.

New barns were supposed to be mechanized. In some places, this helped speed things along, but as one American official viewing a new mechanized pig farm in 1955 noted, "It did not appear to be a highly efficient organization. The mechanized or automated operations, such as the preparation of hog feed, were eclipsed by the amount of hand labor which both preceded and followed the mechanized portion."[13] The official estimated that by mechanizing, Soviet farms had actually increased the amount of human labor needed for their operations.

The other major environmental change took place away from the barnyard, as the Soviet Union began to grow crops for fodder. The heart and soul of this project was establishing field corn as a major new fodder crop. Originally intended as a feed for cows that would replace hay, corn quickly became the feed of choice for pigs. After a visit by a U.S. delegation from Iowa and by other U.S. farmers over the summer of 1955, corn became the centerpiece of Khrushchev's efforts to raise meat and milk productivity. These efforts were what earned Khrushchev his

nickname of *kukuruznik,* or "corn fanatic."[14] Since so little of the
Soviet Union resembled the plains and hills of Iowa, adopting
corn might seem quixotic, but this was a potentially practical
move for a cold country. Unlike the other major fodder crops,
turnips and potatoes, corn could be harvested quite early and
still maintain a high level of protein. It also provided a "gap
month" green feed during July and August, when grazing animals had eaten the first spring growth but the plants had not
yet recovered their biomass. What corn remained in the fields
in late summer was harvested and made into silage, and corn
made the best silage that had been historically available in the
Soviet Union. The high protein content of even silage made
from green mass and unripe corn ears prevented pigs from
losing weight in the winter.

Thus the desire to put a new kind of meat on the table—a
desire first prompted by American food donations of surplus
meats from Iowa farmers adapting to agro-industrial reorder-
ing in their own country—influenced the commodity supply
network of the Soviet Union. Wartime rations were well adapted
to the uncertainty and poor infrastructure not just of war but
of Soviet peacetime too, and these became a source of inspira-
tion for Soviet planners striving to improve the diets of citizens
as well as to shore up the modern and well-ordered reputation
of Soviet agricultural industries. To do this, they ordered up
more and better animals, inventing breeds and paying attention, for the first time, to the efficiency and speed with which
animals were ready to become meat. Reinventing Soviet pigs
had repercussions still further back along the supply network,
and inspired agricultural economists and state planners to em-
brace new farm organizational structures, so that pigs spent
more time inside eating, and led lives of rigid compartmen-
talization that mimicked emerging trends in human urban

society. Beyond the barnyard, a new concern with feed-to-weight conversions led agriculturalists to seek new crops like corn, that were costly to grow but were a perfect food for a pig destined for a *tushonka* tin. Thus in Soviet industrialization, pigs had evolved. They were no longer recyclers of waste items, but rather consumers in their own right; their newly crafted genetic compositions demanded ever more technical feed sources in order to maximize their own productivity.

Soviet ice cream provides a second, apt illustration of how the Soviet Union made available an animal by-product that was hard to produce and technically difficult to supply. Unlike *tushonka* and other preserved meat products, ice cream was a relatively fragile item, and the distribution system that made it available to Soviet consumers reveals some of the flexibility and creativity in the Soviet food system—two adjectives that are not commonly applied to this sector of Soviet industry.

In the postwar era, Soviet citizens were used to standing in line for staples like bread and oil, but in the 1950s a new kind of line started to form. These were the queues for newly available everyday luxuries like ice cream, chocolate bars, and cognac; treats intended to nourish the spirit as well as the body. In the words of one dairy industry specialist in 1961, "Ice cream is beloved by all, and because of this it should become a mass-produced food product, included in the menus at breakfast, lunch and dinner." Soviet food technologists, not necessarily famous for having their fingers on the pulse of popular trends, may have been close to the truth when they observed that Soviet citizens loved ice cream. However, getting ice cream to the socialist masses was tricky. The unusual history of ice cream's production, distribution, and consumption sheds light on the creative Soviet approach in getting animal products into the everyday diets of citizens. Beginning in 1951, ice cream, once a rare luxury of the aristocracy, began to be an affordable product in every major city across the Soviet Union.[15]

In a country that had never succeeded in producing a reliable, year-round supply of fresh dairy products and where refined sugar was available only sporadically, high-quality ice cream made out of precisely these ingredients became a cheap and consistently available treat soon after the end of the Second World War. How did ice cream attain such an important place in the planning agendas of Soviet food distributors? Even more remarkable, Soviet ice cream was a high-tech and high-quality product in a country where many experiments with new domestic technologies were stories of plagiarism, deficiency, and fiasco. What made the ice cream line, stretching not just from vendor to consumer, but also back to producers, food industry planners, packers, and transportation authorities, succeed so brilliantly?

A part of the answer to this question begins with the year 1932, when the Moscow's Municipal Cooler No. 2 installed an ice cream plant and processed its first twenty tons of the unaccustomed luxury food. The rapidly expanding Moscow plant manufactured seventeen thousand tons per annum by the end of 1936. Distribution was initially limited by a lack of freezers and refrigerated transportation, but ice cream was also limited by an unrealistic socialist vision of provisioning popular in the 1930s. Socialist food planners initially believed ice cream should be produced primarily for consumption in communal settings: worker cafeterias, school lunchrooms, and the like. This was consistent with a 1930s push to discourage private family meals in favor of public dining experiences, which in turn was part of a larger movement to create public spaces of ritual and sociality that would replace and improve upon the isolated and potentially bourgeois private home setting.[16]

Ice cream suffered a hard blow during the war, since most of the cows the Soviet Union lost were purebred, high-producing dairy cattle that had lived in the European parts of the Soviet Union occupied by German forces. The cows that survived gave less milk, and all dairy production, not just ice cream, suffered for years. In fact, sugar, milk, and cream were all in short supply well into the 1950s. How could the state make ice cream available to its citizens if these basic ingredients were not at hand?

The answer to this question lies partly in the difference between socialist food collection networks and their capitalist counterparts. While it was difficult for private consumers to access scarce food, it was much easier for the state to do so. The answer also lies in the realm of food distribution and processing technologies. Many processed foods (although not, initially, ice cream) had advantages of distribution over fresh products because they were more shelf-stable and could stand up better to the abuses of shipping and the uncertainties of socialist boom-bust production cycles. It was both easier and ideologically more desirable for the Soviet food industry to create processed foods than it was for it to distribute fresh ingredients.

Cow's milk was the first essential ingredient for Soviet ice cream. Supplies were sporadic and seasonal, but its distribution was also hampered by a lack of refrigeration capacity. Just like potatoes, milk was a commodity that Soviet officials soon learned to collect efficiently, in spite of significant logistical hurdles. Redistributing fresh milk once the state had amassed it was another story. The collection process began on the farm. Cows gave most of their milk in the summer, when it most needed cooling. Fresh milk was first chilled in underground tanks down to ground temperature and held a few days until it was collected.

Dairy products that moved through Kharkov Oblast in eastern Ukraine serve as one example of how distribution worked. Dairy processing plants located in the cities gathered milk from all over the region, but the reverse movement—how far processed dairy products traveled out again—was not nearly so robust. The most fragile product, bottled fresh milk, did not travel very far, and its distribution closely followed the main train tracks that ran through the surrounding districts. The *smetana,* or sour cream zone, stretched further out, several dozen kilometers past every train stop, presumably limited principally by how far light trucks were willing to travel to pick up fresh shipments. Most durable of the processed dairy products was butter, which was shipped out from Kharkov to all regions of the oblast. It's not clear where ice cream fitted in to this reallocation, but it seems safe to conclude ice cream was available primarily to city residents who did not live too far away from the central distribution point.[17]

Sugar was also hard to obtain, and this scarcity reflected a vicious circle common to many Soviet consumer products: there was an unpredictable supply and this led to scarcity. When it was available, consumers hoarded what they could get, which resulted in shortages. State agencies, aware of these practices, did everything in their power to regulate how much sugar consumers could obtain, but had limited control over buyer behavior. Sugar also was expensive to make and buy. The climate of the USSR was ideal for sugar beets and not sugarcane, and of the two products, beets yield a less sweet and more expensive product. While sugar was rare and costly for both state industries and private consumers, the Soviet state had the power to requisition it for industrial use. After repeated attempts to make sugar available in stores year-round, distributing agents seemingly gave up on the year-round policy. Instead the Min-

istry of Food Provisioning released refined sugar for purchase
by private citizens only a few times a year, most notably during
the summer fruit-canning season and over the New Year's and
Easter holidays. This policy ensured that sugar was available
year-round for industrial food production. Instead of making sugar a staple ingredient for home kitchens, the ministry
upped its production of sweetened processed foods, including
boiled candies, chocolate, condensed milk, and, of course, ice
cream. As planners saw it, state-produced goods could better
sate the Soviet sweet tooth than home cooking.

Sugar, milk, and milk by-products posed challenges for
the state because they took extra effort for the country to produce. Sweet and fatty foods were in short supply in the postwar
period, the state worked hard to increase access to these small
tastes of luxury. In the late 1950s, Khrushchev's agricultural
ministers privileged the production of animal products and
sugar to the exclusion of more allegedly practical staples like
grains and root crops. An increased production of cold-hardy
cereals and tubers might have maximized the caloric productivity of the Soviet Union, but instead of focusing on the quantity of staples, food and agriculture officials focused on increasing the quality base of the nation's food supply: its sweetness and
its richness. In the postwar Soviet Union the state wagered that
political legitimacy and social stability would derive at least in
part from the access it could provide to everyday luxuries like
ice cream.

Although this turn toward everyday luxuries depended
in part on a shift in agricultural policy, the success of ice cream
was also dependent on new and better food-processing technologies, especially innovations in industrial refrigeration.
Throughout its history, the Soviet Union had relied more on

mimicry than invention, and many of its food-processing technologies were derived from American or German precursors.
The technologies that dehydrated massive quantities of milk,
for example, came from American equipment developed to
help supply Second World War soldiers with lightweight foods
that could survive harsh field conditions.[18] Contrary to this
general pattern, refrigeration technology in the Soviet Union
was a homegrown innovation. The Soviet Union's research and
development into refrigeration displayed an impressive range
of creativity and adaptability, two traits not often associated
with Soviet engineering.[19]

The expansion of dry ice manufacturing was also crucial
to the development of the ice cream industry. A few dry ice
factories had existed before the war, but the material became
a practical solution for chilling industrial meat lockers and
other municipal coolers during and after the war, since even
at these centralized locations freezers were expensive to operate, noisy enough to elicit complaints from workers, and prone
to breaking down. The war also disrupted central power supplies, and most lockers and municipal freezers that depended
on electric compressors were not equipped to handle a black-
out for more than a few days without major losses. City coolers initially constructed dry ice plants as a stopgap measure to
supply the material for on-site refrigeration in blackouts and
also to use it as a desiccating and cooling agent for the coolers themselves, which tended to overheat and produce excess
condensation.[20]

The role that dry ice would play in uncoupling Soviet
frozen foods from unreliable Soviet freezers went unrealized
for a decade. It was not until 1955 that dry ice facilities received
positive attention from state planners trying to increase the

reliability and productivity of municipal freezers, and it was from this time forward that Soviet food technologists began to think of dry ice as a substance with possibilities.

Ironically, it was a low-tech innovation in distribution that made dry ice popular. By the late 1950s freezers with dry ice facilities had also acquired a fleet of insulated pushcarts cooled by the material. It is not clear what ministry or department ordered these carts, but in the later postwar period they rapidly became a part of the delivery fleet associated with meat- and milk-processing plants. Almost as soon as manually operated pushcarts appeared on Soviet city streets, they were used not simply as retail delivery vehicles; they also dealt in direct sales. Municipal coolers hired vendors to push the carts around the streets during the day. These vendors, typically women, were required to wear coats and white head kerchiefs and follow a strict culture of hygiene. During the 1960s, the "pushcart lady" became an identifiable labor category in Soviet society.

Before the advent of the pushcart ladies, direct sales from processor to consumer were associated with gray areas of private trade. For municipal coolers to employ a cadre of uniformed, hygienically robed pushcart operators to take to the streets as representatives of the coolers themselves was a serendipitous marketing maneuver made possible by the unfettered nature of dry ice. Convenient, public, and associated with a large, well-funded municipal industry, it is possible that ice cream was a popular product not just because it was delicious to eat, but also because, unlike almost every other good in Soviet consumer society, it was fun to buy.

Ice cream sold from pushcarts was most often intended for immediate public consumption rather than home use. The Soviet Union did make "family" sizes of ice cream that could have fit into home freezers, but the home freezer was a domestic appliance whose time had not yet come. Because of both cost and a technology lag, the Soviet Union was slow to switch from ammonia-cooled refrigerators to ones that ran on Freon, and Soviet ammonia fridge-freezer combination boxes had a host of problems: they were loud; their freezers did not get food very cold; water had a tendency to condense on their outsides, hastening corrosion; and their temperature sensors did not always work: a review of the performance of refrigerators and freezers in 1960 noted that "every third or fourth freezer has a serious thermostat defect."[21] The most common problem was a hyperactive relay contact, which led to frequent and messy defrosting events.

Perhaps because there were still kinks in the technology, home refrigerator-freezer combination units were in very limited supply and refrigerator units sans freezer box became the norm for the first-time buyer. Until 1960, the most commonly sold refrigerator model was the *Okean*, or "Ocean," which—aptly named—did not freeze.[22] Thus when Soviet citizens lined up to buy ice cream from pushcart ladies, their most common purchases were individually wrapped cones, popsicles, cups, or sandwiches which were intended to be eaten on the spot. Family ice cream packages rated a distant fifth place in units of sale for the mid-1950s.

The Soviet Union's home freezer technology may have left much to be desired, but the Soviet Union's research and development teams built state-of-the-art industrial ice cream manufacturing equipment. Inspiring this research was a fascination common to many food technology agendas of the postwar period: creating completely processed foods that no human hand touched from start to finish. Soviet food planners often mentioned the desirable hygienic properties of such processing as well as the labor-saving aspect of freeing people from assembly-line work, but the driving force behind much of this preoccupation with total automation was a technocratic fascination with the potential that machines offered for perfect control and infallible order. In the words of one 1960s-era journal: "[In the future] the principal ice cream varieties, sandwiches, cones, and Eskimos [ice cream bars], will be manufactured by continuous, mechanized processes, equipped with modern control and measuring instruments."[23]

Much of the technological research into ice cream production in the late 1950s and the 1960s went into developing fully automated manufacturing processes for the ice cream products (mostly bars and cones) that did not yet have them. The invention of the "Eskimo-Generator" in 1959 brought the Soviet Union ice cream industry closer to its ideal of total automation. As its name implies, the Eskimo-Generator made the chocolate-covered popsicles commonly known in North America as Eskimo Pies (a trademarked brand name held by Russell Stover until 1999), or simply as Eskimos in the Soviet Union. The Eskimo-Generator assembled an Eskimo popsicle from its various components at a rate faster than any human work squads. The Eskimo was the first fully automated ice cream bar, and because of this, it became one of the leading products of the Soviet ice cream industry. The industry steadily invested in automation; by 1970, one ice cream trade journal estimated that the ice cream industry was no less than 80 percent automated.[24]

Although usually made from high-quality ingredients with a minimum of stabilizers and additives, ice cream was not immune from the Soviet state's occasionally misguided ambitions for social improvement, and a line of vegetable-enriched ice creams was introduced in the 1950s. Heralded as a way to get children to eat more beets, carrots, and tomatoes, and given romantic names like "Golden Autumn," vegetable ice cream did not catch on, and these flavors were not produced in great quantities after the early 1960s.[25]

Adding vegetable pulp was an extreme way of establishing the health benefits of eating ice cream; typically promoters simply extolled the virtues of regular ice cream's fattening qualities. To Soviet publicists, ice cream was healthy because of —and not in spite of—its high fat and sugar content. The same 1961 publication that advocated serving ice cream at breakfast, lunch, and dinner also professed that "ice cream is one of the most healthy and delicious food products. . . . both ice milk and ice cream contain fat, protein, carbohydrates, mineral salts, and vitamins. The calorie content of regular ice cream averages 1,450 per kilogram, for high-fat ice cream 1,950, for *plombir* 2,400, and for fruit sherbet 1,230–1,430. It is higher in calories than . . . many other products." If high-calorie products were the definition of health food in the Soviet postwar period, then Soviet consumers tried to eat as healthily as possible. In spite of the plethora of choices available to consumers, most ice cream purchasers had just one of two favorites: crème brûlée, flavored with caramelized condensed milk, or *plombir*, a 15 percent–fat delicacy, with the unappetizing English translation of "putty."[26]

The widespread availability of ice cream in the postwar period gave the Soviet state a flagship product that showed off several triumphs in its new network of socialist food production. Ice cream, previously a rare and seasonal product, was now available year-round, a fact that indicated the success of developing freezer technology and the efficiency of the frozen food distribution system. Soviet ice cream was also made out of high-quality ingredients, which signaled a major accomplishment for a food industry that had previously tried to improve mass distribution by lowering the quality of basic staples.

Even though its quality standards were very high, post-

war ice cream was an inexpensive treat, and it was this economic accessibility that helped to make life a little sweeter for
a diverse group of Soviet urban citizens. Buying ice cream on
the street for just a few kopeks per package was an act of con-
sumerism in which even children could participate, and which
everyone could enjoy. In a country with very little to purchase
recreationally, ice cream was affordable to all. The act of obtaining ice cream was a public, social event, and the Eskimos
and other frozen concoctions were eaten on the spot where they
were purchased from pushcarts: parks, playgrounds, busy street
underpasses, and the courtyards of large apartment houses.

The pathways of production, processing, distribution,
and consumption of ice cream in the postwar era were unique
responses to a challenging food situation. Soviet agricultural
authorities' priorities worked to increase the production of
higher-input "luxury" meat and milk products as well as sugar
at a time when a back-to-basics approach to agricultural sustainability might have been more rational. The state correctly
banked on the idea that the social and political impact of making a few food luxuries available would do more to persuade
a generation of war-weary Soviet citizens of the legitimacy of
the postwar regime than might a focus on virtuous cereals and
root crops.

Another approach Soviet planners took to increasing the
availability of animal protein was to exploit the country's wild
stocks of fish, since preserved fish was a familiar commodity
with an existing distribution network.[27] Dried, smoked, canned,
and salted fish (largely salmon and herring) had been products
of Russian coastlines for centuries. These products had started
out as outpost foods, accompanying Russian expeditions into
Siberia and North America during the reign of Catherine the
Great. The fish-canning industry—one of the few robust food-

processing industries the Soviet Union inherited from imperial Russia—was well developed on both the European and the
Asian coasts by the late nineteenth century and its principle
products were canned crab, salmon, and herring, although dozens of other varieties of canned fish were also available. Many
of these products, like caviar, were earmarked for export as
luxury goods. Although in theory the railroad allowed canned
food products to travel across the continent from either shore,
canned, ocean-caught fish were not distributed nationwide
until the Second World War. Instead, Soviet citizens living
in inland areas relied on supplies of freshwater fish, usually
smoked or dried. This changed once the state made producing meat and milk a priority. The Soviet Union simultaneously
increased its ocean catch rates and quantities, and fish canning
facilities improved in efficiency to the point where canned
ocean fish was usually cheaper to pack and ship than freshwater fish was to smoke or dry.

Immediately after the Second World War three-quarters of
all fish were caught inland and one-quarter was ocean-caught.
By 1959 the ratios had reversed: two-thirds were ocean-caught
and one-third was from inland sources.[28] Fish were one of the
first industries the Soviets actually succeeded in making more
productive by decreasing human labor and increasing machine labor; while this was a goal in all sectors of food production, it rarely worked as planned. By 1959, one-third of Soviet
meat consumption was estimated to come from fish sources,
and the vast majority of these were wild-caught rather than
farmed.[29]

Historians tend to dismiss the role that wild foods have
played in sustaining the Soviet Union, treating it as a crutch that
boosted food production rates. However, the Soviet Union is
not the only developing country to have maximized such re-

sources. In the United States and Western Europe they were largely depleted well before the twentieth century. Hunting, fishing, and gathering are viewed in capitalist agricultural societies as low-input activities meant to complement the work of farming. In a place like the Soviet Union that held so much of its wealth in nonarable land and water, it made good sense to spend extra time and energy exploiting these resources. Though the Soviet Ministry of Agriculture was slow to accept and formally manage nonagricultural food-producing activities like hunting, herding, and fishing, it expanded its economic investments in fishing and sheep herding in the 1950s and, as we shall see in the next chapter, by 1960 the Ministry of Agriculture had redefined hunting as an agricultural activity that fell within its purview.

The human population was not the only group of omnivores to profit from the exploitation of wild foods. In 1961 a group of American agricultural experts visiting the Soviet Union noted that the Belaia Dacha sovkhoz, a large pig farm outside of Moscow, fed its pigs a mixture of feed concentrate, city cafeteria food scraps, and brewery waste, along with the salted heads, flippers, and claws of seals that had been harvested from the Caspian Sea. The American guests were unimpressed with the quality of the seal meat, rating it at "about fertilizer level" and noting that the system of recycling city food wastes "bore little resemblance to restaurant wastes fed to hogs in the United States, a large part of [which] consist[s] of cabbage." In response to questions, representatives from Belaia Dacha stated that the farm processed 60,000 pigs per year. Feeding pigs on the food scraps from factories and large-scale kitchens was a common method of recycling these by-products. The rate of scrap to feed, in this case, two to three, was high for a large enterprise. Even in its waste products, United States agriculture eclipsed that of the Soviet Union, at least in the eyes of these visitors.[30]

Soviet farms may have been exploiting some wild foods, but the potential to take advantage of even more wild resources seemed almost endless. A 1956 memo to the minister of agriculture deplored the underexploitation of miscellaneous sources of protein like small fish, walruses, shorebird eggs, and whales in the Soviet Far East, and suggested that feasibility studies should be done on the economic profitability of gathering such items.[31] Much like the Virgin Lands project, which sought to utilize the unplowed lands of the Soviet steppe, The Ministry of Agriculture's plans to utilize wild animals as a significant protein source shows a scavenger's approach to finding food, one that used the country's natural wealth in wild foods and primary soil fertility to subsidize the short-term goals of agriculture and provide a temporary boost to agricultural productivity.

Beyond meat sources both wild and tame, the malnutrition and repetitive nature of the Soviet diet were worrying to Soviet food planners. In response, the state took steps to ensure that food supplies in general became more varied, plentiful, and nourishing. Food was classified by its fat, sugar, protein, and vitamin content, and generally speaking, the more of these, the better. The 1952 *Kniga o vkusnoi i zdorovoi pishche* (The Book of Delicious, Healthy Food), a glossy cookbook put out jointly by the ministries of Health and Food Provisioning, is an example of this modern food aesthetic in print form. Color photographs feature main ingredients (often canned fish, crab, or beef) rather than completed dishes. The roast suckling pig in the frontispiece casts doubt on whether or not this book had much to do with daily life in the Soviet Union.

While cookbooks such as these often departed from reality, they also offer important insight into the ways in which Soviet eating patterns changed above and beyond the increasing availability of animal products and by-products. Soviet citizens began eating new foods and more varied foods. Some foods adapted better to processes of industrialization than others, and these flourished during this period. Rather than expecting an absolute increase in consumption, Soviet planners envisioned that Soviet citizens would increase their consumption of meat and milk, while decreasing their consumption of less nourishing staples like bread and grain. One food industry expert, writing in 1958, explained, "it should be stressed that as the output of the basic foodstuffs increases, the output of certain other foods will grow less quickly and will even decrease. This chiefly concerns the bread, alcohol, and salt industries . . . the more meat and butter, milk and sugar the population receives the less bread and other food poorer in calories and other nourishing qualities it will consume."[32]

Meat and milk products were not the only food items that inspired the state during the Khrushchev era. Many fruits and some vegetables were well adapted to industrial processes such as canning and drying (the lack of home freezers meant that frozen foods were not popular until the 1970s and beyond). Foods that lost out in the process of industrialization were those that were delicate or perishable. These less industrially robust fruits and vegetables remained seasonal products that remained privately produced. Canned foods that appeared most regularly were those that stood up well to the intensive heat and pressure of the canning process. They were also those foods that had high vitamin and mineral contents, and that could bolster the vitamin nutrition and diet diversity of Soviet citizens. Tomatoes were one such industrial winner. High in vitamins C and A, they were loaded with the vitamin and mineral properties of which any nutritional expert could approve. Tomato juice had been used in the 1930s as an anti-scorbutic, and this medical history made the tomato all the more appealing in a postwar country still struggling to overcome malnutrition.

The rise of the tomato can be charted through the cookbooks of the times. Tomatoes had been cultivated in Russia and the early Soviet Union ever since their introduction in the Russian empire in the eighteenth century, but they were hardly an economically important crop, and were mostly consumed fresh. In the *Book of Delicious, Healthy Food,* a column is dedicated to describing new varieties of tomatoes. But still in this 1952 publication, there are no recipes that call for canned whole tomatoes—only tomato paste, ketchup, and juice. The cookbook has three times more recipes for eggplants than tomatoes, indicating, perhaps, that the former were more widely available at this time, or perhaps just that the Soviet government felt consumers needed more guidance to properly dispatch an eggplant. The complete absence of stewed tomatoes in this encyclopedic cookbook does seem to indicate that such an ingredient may not have been produced by the Soviet state.[33]

However, by 1955 stewed tomatoes were on the rise. Tomatoes had always possessed a kind of theoretical prestige—the *Book of Delicious, Healthy Food* described tomatoes in 1952 as vegetables with "high vitamin content, mineral salts, and a harmonious combined taste of sweet and salty," and claimed that "they have earned a place as the most healthy and delicious vegetable." By 1955 this admiration was translated into an increasing availability of canned tomatoes in shops. Most were produced in the southern republics of Ukraine, Georgia, and Moldova, which had Mediterranean climates and abundant agricultural labor, as all of these crops were hand-harvested and thus labor-intensive. Delicate vegetables including spinach and fresh peas were also canned in larger quantities during the 1950s, but the canning process generally made these vegetables mushy and unappealing, and while their consumption increased, they did not experience the rise from obscurity to popularity of tomatoes, and also peppers. Sturdy, fibrous vegetables that held up well under the heat and pressure of the canning process, tomatoes and peppers were winners in the shift to industrial food. During the 1950s, both became commonly available year-round thanks to new canning technology and a zealous state Ministry of Food Provisioning, eager to rapidly expand into commodity markets it was actually good at provisioning.[34]

The year-round availability of vitamin-rich foods like tomatoes and peppers was the most significant result of the increase in canned food. In Russia and the Soviet Union, the seasonality of diets had endured until after the Second World War because of the low level of food processing. No longer limited by what was seasonally available (which in winter, might be precious little other than potatoes, beets, and cabbages), everyday Soviet diets benefited from canning and other methods of preserving that introduced a wider range of foods.

While many historians of the Soviet Union have focused on the increasing availability of consumer goods in the postwar period, few have focused on the kind or quality of these goods, other than to notice their inferiority and sporadic, urban-biased distribution patterns.[35] Ice cream and *tushonka*, while touted as national foods, were in fact really only available in urban areas. For *tushonka* especially, the urban bias of Soviet planners, extended to distribution networks, made processed meats a de facto city food. Postwar rural farm workers and the residents of small towns had access to a steady supply of fresh pork from under-the-table piglets, but above-board, "legal" pigs rarely made the round trip from farm to factory and back again.[36]

In addition to reforming production, Soviet planners altered consumption trends in food in the postwar period, improving the quality of foods for sale while simultaneously restricting their access. While the pace of postwar consumption increased, Soviet planners took significant steps to slow this process down as much as possible.

Not all architectural designs for places where foods were sold, stored, and prepared were new in the 1950s but most of them were recent. The way places are laid out affects what it is possible to do within them, and the new architecture of markets, apartments, and stores influenced what kinds of foods were bought and sold, how these transactions took place, and how food was stored in shops and in homes. The link between architecture and the possibilities and limitations of an industrial food system hold true also when one thinks of the organizational links within it as one form of architecture. The political and bureaucratic structure of a system influences what is and is not possible. By considering the visible and invisible architecture of a food system simultaneously, it becomes possible to see how both material realities (windows, power outlets, refrigerated cases) and the political and economic infrastructure that construct food systems make certain things easy while making other things impossible.

In the 1950s, an accessible architecture began to emerge in the Soviet Union, one designed around socialist notions of access. The birth and development of a culture of ready-made food was constrained by cultural norms, the architecture of commercial spaces, and the presence or absence of home appliances and storage spaces for food. Regarding cultural norms, it is important to understand traditional eating patterns in the Soviet Union. As we have seen, the Soviet diet, especially the diet of urban workers, became more varied and less repetitive in the postwar period. While food had periodically been in short supply, food available in cities had never before been as varied as it was after the war. It was often neither affordable nor plentiful, but the number of different kinds of foods being offered for sale had never been greater. Many of these new products seemed to be geared toward the busy schedules of dual-career working families. In 1959, there were dozens of different kinds of meat preserves and frozen meatballs, many with names like *Pionerskii* (the name of the children's scouting organization), *Maliutka* ("Little One"), and *Shkol'nye* ("School-style"). The Soviet Union did not have consumer research that indicated processed meats would appeal to busy working mothers, but the food product names indicate that this was one way in which these foods were marketed.

In the postwar period, more Soviet citizens were eating more meals in canteens and cafeterias than ever before. Khrushchev created single-family apartment dwellings and phased out communal apartments, which signaled a concession to the right of families to cultivate and spend time in a private domestic sphere, but Soviet citizens were still encouraged to spend as much time as possible in public places. Although restaurants were intended for tourists and travelers, workers' cafeterias became important sites where the state could display both its competence in preparing healthy food and its ability to one-up home cooks by serving foods such as fruit cocktails and cuts of fresh meat that citizens rarely had access to in the general marketplace. Thus, through worker's cafeterias, food norms changed. While food prepared for the masses rarely left much room for culinary artistry, the better access state kitchens had to a range of ingredients meant that cafeterias offered a wider variety of dishes than those available at home. Cafeterias changed eating patterns by anchoring the schoolchildren's and workers' day around a hot lunchtime meal.

Related to the success of the Soviet cafeteria was the decline, or continuing neglect, of the home kitchen. Apartment kitchens in communal spaces were often unusable. At best, a family living in a communal apartment had a two-burner stove to call its own, a shared sink for washing up, and a dish storage cabinet. In reality, the clutter and confusion of multiple families sharing a single room to prepare separate meals often rendered communal kitchen spaces dirty, crowded, and smelly. Things improved after the collective model was abandoned. Postwar urban architecture focused on modest modern single-family apartments. These retained the traditional design feature of a niche or breakfast nook that served both as extra counter space and as a space for drinking tea and informal visiting. Ideally, new kitchens came with appliances that made food storage and preparation easier. The typical postwar apartment had a deep sink, a small breakfast table with bench seating, a dish cabinet, and a Hoosier or other multipurpose cabinet.[37] Kitchens remained very basic, since most families could not afford elaborate furniture or, especially, a luxury like a refrigerator. Families lucky enough to get apartments with balconies were often able to convert at least part of their balcony space into a modified cold pantry, but kitchens were built without pantries. Sale catalogs of kitchen furniture offered bins to store ten or twenty kilos of root vegetables, modest amounts by the standards of the day.[38] The architecture of new kitchens dictated the use of the space, namely, making hoarding or long-term food storage difficult or impossible, and necessitating frequent trips to the store. Urban residents, already limited by their amount of free time and by what was available to buy at stores, were further limited by what it was possible to cart home from the store, and where they could keep such things once delivered.

Much like the architecture of new homes, that of stores limited what shopkeepers could purvey. The minimum technical requirements for stores, unlike those for apartments, dictated that stores should have pantries or dry cellars in which to store bulk root vegetables and other items like flour and fruits. Most stores did not have refrigerated coolers; they relied instead on ventilation to keep foods relatively cool and fresh. Soviet engineers and planners worried quite a bit about achieving good air circulation in stores and storage spaces. The Soviet Union produced manuals and numerous instruction charts on how and when shopkeepers should open windows and doors, and how they should position various stock products around the store and the storeroom in order to maximize air flow, thereby (in the eyes of Soviet scientists) guaranteeing the health and freshness of the food to be sold.

More limiting than the primitive state of preservation technologies at grocery stores was the awkward way in which items were sold, which was often described with disdainful incredulity by visitors to the Soviet Union who were used to self-service–style American shopping. The process involved standing in one line and placing an order for a particular item. The clerk taking the order made out a bill, which the consumer then settled at a central register or more commonly, since cash registers were scarce, with a clerk holding a cashbox and abacus. The consumer was issued a proof of purchase ticket. With this ticket, the consumer then returned to the sales area, where the original clerk handed the purchase over and tore or marked the receipt to indicate a complete transaction. Thus each purchase actually involved three separate interactions with either two or three store clerks. Compared to the popular self-serve food markets of the United States, the process seemed arcane and ungainly.

The setup was meant to accomplish several tasks that are more important in an economy of scarcity than they are in a demand economy. The first of these was discouraging shoplifting and other kinds of stealing. While various forms of graft and embezzling were common throughout the Soviet period, store purchases were monitored closely in an attempt to deter stealing either by consumers, or by shop clerks. Having several employees accountable to each other as well as to customers made it much more difficult to directly bribe them or for them to quietly remove store inventory. This seems to have mostly just resulted in more creative forms of theft, rather than actually discouraging illegal practices.

Another reason for the bulky purchasing arrangement, at least in the case of food, had much to do with the fact that the Soviet Union was not really set up for visually displaying foods or other products. Problems of production and distribution plagued the country generally, but packaging of food proved to be a particular Achilles heel that took decades to overcome. Even if consumers had been able to see what was in coolers and on shelves, they might not have been able to recognize it. Without plastic packages, with a shortage of jars, and with inferior labels that fell off or were unreadable at a distance, the Soviet Union's food supply was simply not engineered for a self-serve populace.

Ultimately, Soviet diets improved but were not revolutionized, and Khrushchev's promise to catch up and overtake the United States in meat and milk production was never realized. The Soviet Union's very public campaign to increase the availability of milk, butter, and meat experienced many successes, and foods like *tushonka* and canned fish supplied a huge boost in protein availability in the 1950s and in the decades beyond. Ice cream made Soviet life just a little bit sweeter, and new convenience foods catered to the limitations of Soviet daily life by offering a wider variety of nutritious options that did not require much time or effort to prepare. The new, improved form of Soviet eating suffered the same urban bias as many other industrial projects. While food was produced in the countryside, it was processed, stored, and distributed within cities. Rural people were obliged to travel into cities to obtain foods other than the basics, and new processed and packaged foods rarely appeared in rural stores.

Nonetheless, for the Soviet Union's increasingly urban population, diets really were changing—mostly, although not entirely, for the better. By 1960, Soviet consumers were eating more meat, dairy, and eggs, as well as more packaged and processed foods, than ever before. Diets were also becoming more varied and less tied to seasonal variation. Some foods that had almost always been consumed fresh just a generation earlier—tomatoes, peaches, and peppers, for example—were now much easier to purchase in cans. Nevertheless, the Ministry of Agriculture's stated goal of surpassing the quality and variety of the American diet was an unrealistic ambition.

Improving Soviet diets was only one objective among several. Another goal that was less trumpeted in the press was that of modernizing food distribution networks that could turn all kinds of foods into normal, popular consumer items instead of chronically scarce, rationed commodities. While Soviet food shops of this era hardly looked like modern, successful enterprises in the 1950s, with their long lines and laborious purchasing rituals, they were in fact institutions of convenience for producers and distributors who wished to limit consumption: they allowed a larger number of people to have access to supplies, but made obtaining them awkward and time-consuming. Convenience, in a socialist food network, was in the eye of the beholder.

Soviet state nutritional policy was committed to two goals simultaneously: first, improving the health of its citizens through diet; and second, making the most out of what was available. To this end, nutritional research was directed at overcoming the long-standing threat of malnutrition as well as making the products that the Ministry of Agriculture could actually procure into new consumer essentials. Cookbooks, food advertisements, and nutritional pamphlets offered practical advice about what citizens could and should eat, and this advice informed the public about good nutrition while at the same time persuading it to adopt distinctively socialist modern diets. Perhaps more than any other material artifact, food reflects culture, so the development of a Soviet pattern of eating should come as no surprise.

5

The Old and the New

Projects for improving Soviet agriculture in the post-war period typically focused on the most productive regions of the country: Ukraine, central Russia, and the Black Earth Region south of Moscow, where agriculture was central to the economy. Here, however, we examine the environmental and geographic limits of agricultural reform in a region where agriculture was historically a secondary pursuit—Siberia, in particular Irkutsk Oblast, far to the east near the border with Mongolia. What kinds of rural and agricultural modernization plans made sense to the state in this frontier region of the Soviet Union, and what sorts of unique challenges did the extreme environment of Siberia pose to reform? How did these plans affect environments and work cultures on the Soviet Union's coldest periphery? When were state interventions successful, and when did they fall short of their stated goals?

To begin with, we look at the history of agriculture and collective farms in Irkutsk Oblast. By the postwar period, many of these farms were struggling. The hardships farmers faced suggest that by the 1950s, agriculture had reached the environmental limits of the region's ecosystem and the Soviet government's options were limited in terms of improving the productive capacity or efficiency of such operations. Nevertheless, a series of new state farms, part of Nikita Khrushchev's much-publicized Virgin Lands Campaign, were planned for the region to ratchet up both grain and fresh fruit and vegetable production. Although agriculture was at the center of Khrushchev's reform policies, in Irkutsk Oblast industrial development dwarfed agricultural development. Analysis of the interplay between an ambitious hydroelectric dam project and the more modest agricultural modernization reforms the Soviet state envisioned for the region between 1955 and 1965 shows that neither set of reforms experienced great success; Irkutsk Oblast was cold and remote, but most damning to these projects were the oblast's idiosyncrasies. Eastern Siberia was a part of the Soviet Union, but it was also a region with different cultural norms, settlement patterns, weather patterns, and environmental limits from those of the western USSR. Central planners, eager to develop and improve the region as rapidly as possible, failed to understand these differences.

We then examine a series of more modest, but significantly more successful state reforms in the region. Beginning in 1960, the Ministry of Agriculture began to sponsor a small number of hunters in the oblast. Hunters were usually also collective farmers, and they primarily hunted sable and squirrel. This was a far more profitable endeavor for both the state and its citizens who participated in the hunting programs, which succeeded largely because the state paid close attention to environmental factors, and because hunting was a traditional endeavor in the region that was well understood by many people. The state managed to successfully insert itself into this economic activity by taking over the role of a relatively

corrupt class of capitalist middlemen who had previously collected furs and paid hunters. In Irkutsk Oblast, hunting helped
the Soviet state redefine its smallest and most marginal farms,
while simultaneously demonstrating that the state could overcome environmental obstacles in order to establish greater political and economic control.

After the Second World War, Siberia remained a region
of benign neglect until the mid-1950s, largely because it was
remote and sparsely populated. This era ended in 1952 when
several Soviet agencies, including the Ministry of Agriculture,
announced ambitious plans for industrial development in Siberia, Kazakhstan, and eastern regions. These projects included
a massive hydroelectric dam, renewed attention to a plan to
extend the Trans-Siberian railway, and new, deep-shaft mines.
The projects would exploit the considerable natural resources
of Siberia while simultaneously giving the state a better control on managing both the resources and the people of the regions concerned.

In 1950 Irkutsk Oblast had a population of fewer than one
million people, with a single major city (Irkutsk itself), a few
dozen collective farms, and very little other industry. Located
far from the lifelines of rivers and roads and dominated by its
extreme cold climate, the oblast was not a self-sustaining—
much less a profitable—agricultural landscape. In spite of its
isolation and low population density, the oblast was included
in the industrial reform plans first announced in 1952, and initiated in 1954. Irkutsk would host one of the most important
development projects in Siberia: the world's largest hydroelectric dam, to be built in the northern part of the oblast across
the powerful Angara River. This project, the Bratsk Territorial
Production Complex, or TPK, was actually a series of massive
construction projects that took place between 1954 and 1971.

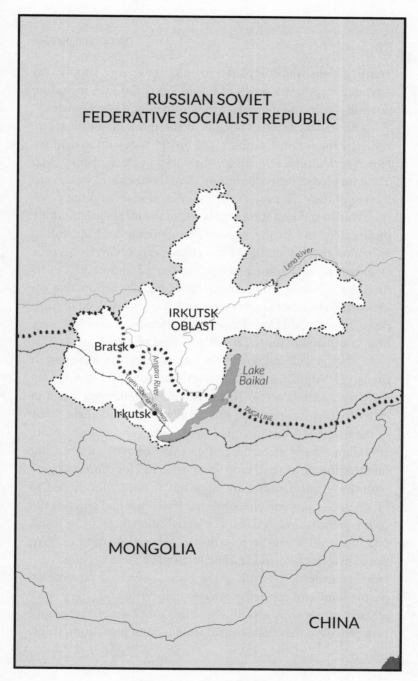

Map 4. Irkutsk Oblast in 1960

The first dam was completed in 1961. Its energy supply was intended to power a series of future industrial sites including sawmills, pulp and paper factories, mines, and greenhouses.

Almost as an afterthought to the more glamorous Territorial Production Complex, the Soviet state announced regional agricultural reforms for the oblast in December of 1954. The dam project called for hundreds of thousands of workers to live on or near the project site for several years and the dam's reservoir would also submerge some of the most agriculturally productive land in the region. Reorganizing and increasing agricultural production for the oblast were about to become essential. In 1954 Irkutsk already imported almost half of its food supply from other, more agriculturally productive regions, and the labor and transportation costs involved in shipping food already meant that food prices in the oblast were high compared to more centrally located regions.[1]

The agricultural reforms announced in 1954 were part of the larger Virgin Lands Campaign that proposed plowing up steppe grasslands across Kazakhstan and southern Siberia in order to grow more grain, specifically wheat and corn. Some of the new grain was intended to provide a cheaper and more abundant supply of bread and cereals to Soviet citizens, but much of the anticipated crop was earmarked for cattle and pigs; with the Virgin Lands Campaign the Soviet Union hoped to finally solve the problem of animal feeds that had plagued the country since the early days of collectivization. In Irkutsk Oblast, one of the most northern territories selected for reform, proposed projects were modest and focused on feeding people, not animals. Indeed, the bulk of the project involved plowing up pastureland and replacing grasses that cattle and sheep were already eating with spring wheat and other crops that would help provision the oblast's anticipated human population surge.

The campaign intended to place some 426,000 hectares of virgin land under cultivation, almost tripling the amount of cropland in the oblast. These plans also sought to replace the rich, alluvial fields that the oblast would lose to the dam project with new grain and vegetable sovkhozes that would be created in its extreme southwest corner, a steppe zone that had traditionally served as pasturelands for sheep, cattle, and horses.

These projects, modest in scope when compared to the whole of the Virgin Lands Campaign, ultimately proved too ambitious for Irkutsk's limited human and environmental resources. However, the state had a second agenda for both its Virgin Lands Campaign and the Bratsk TPK which was ultimately much more successful—namely, to encourage a second wave of European colonization throughout Soviet Asia. Beyond their status as development initiatives and economic ventures, the Virgin Lands Campaign and the TPKs were geopolitical initiatives that sought to place more of the eastern Soviet Union's resources at the disposal of the Soviet state.[2] The state had great ambitions for the TPKs, but the Ministry of Agriculture was under no illusions that Irkutsk Oblast could be transformed into a highly productive agricultural landscape of the best quality. Instead, the goal in Irkutsk was to minimize the costs associated with provisioning a settled, European population of industrial workers (and some agriculturalists) in a remote territory that was, nonetheless, economically and geopolitically vital to the larger nation.

Previous European settlement in Siberia had been limited to river valleys and railway main lines. The settlers there rarely ventured further inland than the transportation corridors they monopolized. Irkutsk was not an obvious center for either farming or industry; just over 3 percent of the land in the oblast was considered arable, and most of the territory was forested. As late as 1950 seminomadic indigenous Buriat shepherds and ranchers made up almost a third of the oblast's population. Because of the industrializing initiatives of the state, the late 1950s and early 1960s were a period of heavy European immigration into Siberia. By the end of 1962, migration, largely from Ukraine and the European parts of Russia, had tripled the population of Irkutsk Oblast from its postwar figures. The Bratsk hydroelectric project alone inspired a miniature boom, as the population in the Bratsk region increased from 55,900 in 1955 to 206,600 in 1966. As anticipated, immigrants increased the demands on the oblast's environment, food supply network, and housing stock. The ambitious development plans for the oblast rapidly exceeded its capacity to provide basic necessities to the expanding population.[3]

Irkutsk Oblast was a territory in which the Soviet Union was willing to invest because of its abundant natural resources and its long history as a center of trade and transportation for the country's Far Eastern territories. Before European settlement, the shores of Lake Baikal and the Angara River had been inhabited since at least the twelfth century by Buriat pastoralists. Most European Russians who passed through had been traders or hunters using the Angara River as a thoroughfare, though the town of Irkutsk had already been important as a layover point for hunting, trade, and mapping expeditions headed farther east. The history of agriculture in the region was significantly shorter; farmers began to settle in the region that would become Irkutsk Oblast only in the late eighteenth century, after the region became part of the Russian Empire. By the nineteenth century, enough Europeans had settled in the region to prompt an informal geographical division between Europeans and Buriats, with European farmers setting up farms to the west of Lake Baikal and Buriat herders keeping their flocks to the south and east of the lake. In general these southern pastures were steppe areas well suited for pasturing sheep and horses. European immigrants to the region established farms along the banks of the Angara and Lena Rivers, and along the shores of Lake Baikal. Near these bodies of water the soil was richer and the climate a bit milder.

Europeans who moved to the Irkutsk region from points west were a motley collection of Old Believers expelled from the Russian Orthodox Church, escaped serfs, convicts exiled to Siberia as punishment for their crimes, and professional traders, called *promyshlenniki,* whose role in the history of the region will appear later on. For the minority of settlers who chose to farm rather than to hunt or trade, a short growing season limited crop choices, but agrarian households still usually raised crops both for their own consumption and to sell for profit; farming in the region was high-risk, but it could also yield high profits. Because of its strong market orientation, Siberian agriculture in this early period did not resemble agriculture in other parts of the Russian Empire; most people who lived and worked in the region depended on a small minority of farmers to produce the bulk of the region's food.

Eighteenth- and nineteenth-century farmers established a unique and enduring rural culture that lasted into the twentieth century. As late as the 1960s, anthropologists noted that Irkutsk Oblast agriculturalists used a different vocabulary, owned different tools, had different farming techniques, and practiced more communally based forms of rural enterprise than their counterparts in European Russia. At the start of the postwar era, over 95 percent of the population of Irkutsk still inhabited the Lena and Angara River valleys. Even in the postwar Soviet Union, riverine farms were small and diverse; a shortage of human and horse power limited both plot size

and productivity. Although most farms had officially been collectivized in the 1930s, mechanization and electrification, the
hallmarks of kolkhoz modernization in the European Soviet
Union, still had not reached the majority of Irkutsk's collective farms in the postwar era. In 1949, the oblast had 38,449
draft horses and fewer than 2,000 low-horsepower tractors,
making horses the most important power supply in Irkutsk
Oblast. Of the 547 kolkhozes in Irkutsk Oblast in 1958, only 140
had electricity. Collective farmers plowed with horses, used
kerosene lamps, and hauled water from wells in old milk cans,
sometimes from as far as five kilometers away. Just as their premodern counterparts had done, Irkutsk Oblast farmers in the
postwar period typically raised a mix of buckwheat, barley,
wheat, and a few vegetables, as well as turnips and hay as animal feed.[4]

At first glance, Irkutsk Oblast in the first decade of the
postwar era was a rural backwater that had changed little from
preindustrial times. However, technological progress had not
completely bypassed the region. The railway, completed in 1891,
was the first major transportation network linking European
Russia to Siberia.[5] The rail line was not always reliable—ice, de-
railments, and engine failures were all too common—but the
Trans-Siberian represented a regular and comparatively timely
link between Siberia and European Russia. Nationally, rail lines
connected Irkutsk to Moscow, and within the oblast a spur
linked the rapidly expanding city of Bratsk in the north to Irkutsk in the south. By the 1950s, the written word also linked
the countryside to cities and towns; in 1951 the oblast published forty-two different local periodicals, at least twelve of
them dedicated to agrarian subjects.[6]

Aside from the region's remote location, soil quality limited agricultural production. Irkutsk's soils, like Kazakhstan's,

were podzol, a loose, acidic, sandy material, prone to erosion and without much clay to hold it together. The soils of the river basins held more sediment and were better for farming, but they were still podzols, which meant they were easily depleted of nutrients and eroded under intensive cultivation. The ancient and inefficient wooden plows and harrows that local farmers continued to use in the postwar era century were better suited for these marginal fields than the heavy steel implements typically associated with agricultural industrialization in the European Soviet Union. Thus, environmental, geographical, and cultural factors all limited agricultural production in Irkutsk Oblast. Premodern techniques and the soil limitations of the region curtailed the productivity and efficiency of farming. Farming was labor-intensive; large-scale or industrial style farming was virtually nonexistent until the Virgin Lands Campaign began in 1955.

The Virgin Lands Campaign aimed to increase the productivity and profitability of agriculture while Territorial Production Complexes targeted heavy industries like mining and construction projects. Both campaigns shared the goal of putting Siberia to work for the Soviet Union in new and more intensive ways. It was only a minority of the oblast's new arrivals who joined the Virgin Lands Campaign; far more workers moved to Irkutsk Oblast to work in the new extractive industrial bases the Territorial Production Complexes created. In Irkutsk Oblast alone, Soviet citizens arrived to work in gold mines, on logging crews, at a new aluminum smelter, and at six different construction sites along the Angara and its tributaries.[7] Of these work opportunities, the largest employer of new immigrants was the Bratsk dam project.

Most migrants into Irkutsk between 1955 and 1965 were young—under thirty. The state actively encouraged Soviet citizens at the start of their working lives to move eastward because European cities were overcrowded and housing for new migrants was critically short. A large number of immigrants into Irkutsk Oblast came from Ukraine, seeking to start over in a territory that had not been ravaged by war. The Ministry of Work advertised its eastern settlements as desirable places to relocate, and offered free train fare to families with two or more full-time workers. In a bid to attract young couples as well as single people of both genders, the Bratsk project actively encouraged women as well as men to sign on for construction labor, and in 1961, approximately one-third of the Bratsk project workers were women. This was not a novel approach by the Soviet state; women had been specifically targeted as ideal candidates for assisting in the settlement and Europeanization of the Soviet Far East since the 1930s. Ads for eastern industrial projects stressed the high quality of life and the variety of jobs available. Moving to Siberia or Kazakhstan was pitched as an adventurous and patriotic choice for young citizens just about to enter the workforce. Sharing in the Soviet Union's drive to industrialize Siberia and Kazakhstan gave Soviet youth an opportunity to participate in history, and also to connect with the experiences of their parents and grandparents, who had participated in great works projects during the First and Second Five Year Plans during the 1920s and 1930s.[8]

There was often considerable distance between the advertised conditions in the eastern Soviet Union and the reality of daily life for new immigrants. Everyday consumer goods such as curtains, glassware, shoes, and clothing were scarce in the region well into the 1960s. On the Bratsk dam construction site workers walked around in "Studebakers," thick homemade shoes with soles cut from worn-out truck tires. On the newly created sovkhozes in Irkutsk Oblast, community centers such as clinics, schools, and bathhouses were either exceedingly rare or nonexistent. Often hundreds of workers shared one bathhouse, which was the only community building with hot water. On construction sites, new immigrants were often housed in earthen dugouts or in tents, temporary solutions to housing shortages that developed into uncomfortable, multiyear dwelling arrangements for thousands of families. Both tents and dugouts were especially uncomfortable in the Siberian climate; tents were stifling hot in the summer and freezing in the winter, earthen shelters were always damp.[9]

During its first five years of construction, the Bratsk hydroelectric dam project employed forty thousand workers around the clock in shifts of nine thousand. One American visitor in 1961 wondered at the motivation of the workers in such a remote location and unpleasant circumstances. His impressions are worth quoting: "The reporting officer questioned . . . how these young people were induced to come to Bratsk with its ninety degree range in temperature and its two summer months marred by the appearance of biting gnats which require workers to wear a protective netting over their head. Answers varied. Some attributed the willingness of Soviet youth to serve as a mark of their high devotion to building communism. Others mentioned the extra pay Bratsk workers received for working in a Northern climate, equal to an additional five percent for each year served."[10]

The Bratsk hydroelectric station was a classic Soviet great works project, breaking several world records in construction statistics. Once fully operational, it nearly doubled all of Siberia's electrical capacity. The planned reservoir for the station was massive—the world's largest for several years—with a nearly 170-square-kilometer surface area. Nine villages and their adjacent farmlands were inundated when the reservoir was filled, and their loss inspired a rare moment of nostalgia among both Soviet writers and social scientists. Valentin Rasputin's classic village novel *Farewell to Matyora* (1976) takes place on an island settlement in the middle of the Angara River during the last spring before the village is flooded. A number of ethnographies that emerged during the 1960s also focused on describing and celebrating a uniquely Siberian agrarian lifestyle that was rapidly disappearing to make way for industrial progress. Liudmila M. Saburova's *Culture and Everyday Life in a Russian Settlement on the Angara* (1967) was the most widely circulated of this nonfictional genre.[11]

The new hydroelectric station also created a typical Soviet cart-before-horse conundrum; the biggest obstacle the project faced was not producing power, but finding consumers for the electricity the project would generate. Irkutsk, which had its own hydroelectric power station, lacked the industrial infrastructure to take advantage of even a tenth of the additional electricity of the planned Bratsk station, and while Soviet engineers excelled at building power stations, they stumbled at effectively and economically distributing the power. Aside from Irkutsk, the nearest population center to Bratsk was Novosibirsk, over twelve hundred kilometers away, a city with its own plentiful hydroelectric power sources. The closest population center actively in need of electricity was Sverdlovsk (now Yekaterinburg) on the far side of the Urals, over five thousand kilometers away.[12]

Developing the hydrological capacity of the Angara by building a dam at Bratsk created more problems than it solved, a fact that would have halted a less visionary project. Not so with the Soviet Central Committee as it drew up the Sixth Five Year Plan for 1956–60. Soviet industry and its economy worked best when there was just such an identified area of underdevelopment that could be seized upon and intensively improved. Experts estimated that creating a fully functional industrial corridor along the Angara would cost more than eighty billion rubles and require an additional 3.2 million workers.[13] The cost of the most famous great works projects of the Soviet era combined—the White Sea Canal, Magnitogorsk, and the Baikal-Amur Main Line—were all dwarfed by this investment in capital and workers. Yet so great was state confidence in its project to industrialize and populate Siberia that the state banked on this improbable future; in both the Sixth and the Seventh Five Year Plans, the dam construction at Bratsk was treated as a foundation for much bigger projects that would result in the permanent occupation and exploitation of a wide variety of natural resources across the entire Siberian subcontinent.

The Bratsk dam project did not just employ immigrants from distant parts of the Soviet Union; it also siphoned farm labor away from Irkutsk Oblast's sovkhozes and kolkhozes. Even with inferior housing and chronic shortages of common consumer goods, the Bratsk project offered better pay, better hours, and better living conditions than regional farms could provide. In their 1962 annual report, the sovkhozes of Irkutsk Oblast collectively predicted a net gain of 663 workers on state farms. Instead, they lost a stunning 18,498 workers that year. Of course, every worker who left a sovkhoz did not join a Bratsk construction project, but many did, and the net loss of sovkhoz workers indicates that these employees did not just change jobs, but also labor sectors. Agriculture lost out. The oblast's kolkhozes were even worse off because of the much smaller scale of their operations. On the tiny, bankrupt farm Bolshevik, for example, sixty-one workers left between 1950 and 1960, leaving only forty full-time workers. During a boom, in remote locations like Irkutsk Oblast, the state had little ability to keep agricultural workers on farms using either persuasion or force; incentives to pursue other kinds of work were simply too powerful.[14]

A major mistake the Ministry of Agriculture made when addressing food provisioning in Siberia was to site farms based on an approach common around European Soviet cities, by artificially manufacturing a series of distribution rings. Specialty crops, taxes, and prices were all planned out in concentric circles, with higher-priced and more fragile commodities located closer to urban centers, and more robust and less valuable commodities produced further from cities. In essence, these were planned von Thünen rings. The intent was to make food production as local as possible, reducing transportation costs and loss due to spoilage or poor coordination. Moscow, the best-realized (and most functional) model of this system, had dairy farms and fruit, potato, and vegetable farms immediately outside the urban district. Further out in a second ring, pig and chicken farms raised meat for the city.[15] One ring past the pig and chicken farms were fruit orchards and grain fields. As the USSR's largest city, Moscow was a sink for produce, meat, and other perishables, and these rings were not robust enough to make it self-sufficient. For a medium-sized Soviet city, the ring system succeeded at earmarking strategic suburban spaces for farming, and the system allowed many Soviet cities to be self-sufficient in food.

These farm rings worked well across the European USSR, and by the 1960s every major city had a reliable state-produced supply that put fresh fruits, potatoes, and milk on sale in cities. The novelty here was not that locally produced foods were being sold in cities, rather that the official state-funded farming system had succeeded in this task. Previously private allotment gardens had supplied most cities with the bulk of their fresh produce. This system of carefully siting farms for cities based on the distance food needed to travel to reach its intended market was one the state had appropriated from capitalist food distribution networks. With effective management and competent planners, there was no reason to believe the system would not work in any location where an urban population lived next to fertile farmland. Irkutsk, however, was not such a place, and this was the detail that Soviet planners overlooked, or chose not to see.

Postwar immigrants to Irkutsk had moved into the region to build a dam and participate in other developing industries, lured by the promises of higher wages, better working conditions, and the opportunity to participate in a project to improve and enrich the country. Settlement patterns in the oblast diverged from those in other parts of the Soviet Union for a variety of reasons, including the harsh climate, the well-established indigenous ranchers in the region when European settlers arrived, and the difficulties of traveling overland during the muddy fall and spring seasons. As a result, Irkutsk's agricultural settlers did not conform to the ring pattern the Soviet state anticipated. Their settlements tended to stretch out in long ribbons that followed the shorelines of Lake Baikal and the significant rivers, but nobody paid any attention to the ribbon pattern that was standard to the region around Irkutsk, and this created problems. Sometimes the ring patterns imposed on the oblast worked, notably the dairy ring, while most others failed, especially the more crucial produce and grain rings. Around the cities of Irkutsk and Bratsk, greenhouses and dairies were placed within a half-day's train ride from the city, much like the "milk shed" that provisioned Ukraine outlined earlier. Most vegetables were raised in greenhouses, and their productivity was not reliable. Greenhouses could stretch

out the growing season by keeping out the cold, but few greenhouses were equipped with electric lighting, and these structures
did nothing to keep out the dark, and the very short spring and
fall days limited seed germination and growth. Although dairies and meat-producing sovkhozes fared better, they were not
completely successful ventures either, at least not at first. These
operations were expected to produce their own winter fodder,
partially in the form of corn, which had become the feed staple
for the rest of the Union. Dairies struggled with the lack of sunlight as well, and were only commercially productive for five
months of the year because of the harsh climate. Dairies and
open-range cattle grazing were the largest expenses for the Irkutsk agricultural economy.[16]

The ring model's failure was especially obvious in proposed areas of intensive industrial development like Bratsk.
In 1960 the city of Bratsk relied for milk, eggs, and vegetables
on just nine small kolkhozes located upstream from the dam
project.[17] The city of Bratsk, a magnet for new immigrants, had
no truly local source of grain; the large grain sovkhozes that
the Virgin Lands Campaign created were all located in the opposite corner of the oblast, at least two hundred kilometers
from the city. The nine kolkhozes that supplied Bratsk with vegetables, milk, and fruits were all scheduled to be flooded when
the reservoir began to fill in 1961, and it would significantly
diminish the fertility of cropland downstream that depended
on the Angara's annual flooding to restore organic material
to fields. The newly established fruit and vegetable sovkhozes
that were intended to replace this lost farmland were not yet
producing enough to make up for the loss of the riverine
farms, and after 1961 the oblast became even more dependent
on outside food shipments. Grain and other staples arrived by
train from European Russia, but oblast residents went without

many more optional foodstuffs for years. Prices for all foods remained high, and unlike many parts of the Soviet Union, private production did not help supplement and make up for shortages at the official markets; there were simply too many workers who had no way to raise food.

When compared to other parts of Siberia that participated in the Virgin Lands Campaign, Irkutsk's agriculture was strikingly marginal. In a memo dated March 1956, an Irkutsk official from the Ministry of Agriculture wrote that the combined granaries in Irkutsk had a holding capacity of 16,000 tons of grain, but only 9,600 tons had been stored by the end of August, the last month of the wheat harvest. In contrast, Omsk Oblast, a neighboring territory with a larger Virgin Lands project, had the capacity to store 128,000 tons and already had 60,800 tons in granaries. Shkalovskaia Oblast had a capacity of 211,000 tons and by the end of July (in the middle of the harvest and when Irkutsk granaries were still empty) already had 70,000 tons. In other words, even in 1956, one of the best years for grain production nationally, Irkutsk Oblast could not meet the modest production goals that it had set for itself. Although it was scheduled to produce only about 10 percent of what its neighboring territories were scheduled to produce, this production figure was not possible for Irkutsk's farms. In 1961 an Irkutsk political official proudly announced to a group of American agricultural officials visiting the region that the oblast was not only grain self-sufficient but also a net exporter of grain. The Americans were suspicious of this statement, and they were right to be so; this was simply not true. Siberia as a whole was self-sufficient in grain by the early 1960s, but the central and eastern oblasts were net importers of basic foodstuffs, including grain.[18]

There is some evidence that Irkutsk's small kolkhozes

were even worse off than its grain farms, most of which were sovkhozes. Because they were state-owned, sovkhozes were also state-subsidized, and their losses were absorbed by grain and cattle trusts. The smaller kolkhozes did not have this kind of insurance policy; instead their failures were directly absorbed by the kolkhoz members in the form of reduced wages and reduced capital funds.

Beyond the weather and the shortage of labor, many ring farms were prone to fail not just because of settlement patterns, but because the economies of the USSR's outpost holdings functioned differently from those in the main part of the country. Historically, Siberia, Kazakhstan, Mongolia, and Manchuria had all enjoyed close ties to the Russian empire and later (with the exception of Manchuria) to the Soviet Union, but these regions functioned more like colonies than fully participatory states; the Russian, and then the Soviet empires adopted the role of benefactor. All of these lands were agriculturally marginal, with poor soil and sparse human populations. The indigenous economies that had endured in Irkutsk and the other northern and eastern provinces were those that relied on livelihoods other than traditional field agriculture: hunting, fishing, and herding sheep or reindeer. The Europeans who had previously settled in Irkutsk had wisely adopted similar lifestyles, supporting themselves through diverse occupations, including, but not predominantly, farming.

Large-scale industrial and agricultural reforms were the economic focus of the state throughout the 1950s, but Irkutsk's true value did not lie in converting its acidic soil and its larch forests into profitable farmland or in harnessing its powerful rivers as electricity sources. Rather, its wealth was stored in its forests, filled with high-quality timber; its soil, which held precious minerals; and its animals, the fur-bearing creatures of the taiga. These animals held the key to one of the most successful reforms the region experienced during the 1960s. Partially as a way to offset the financial drain of the struggling new sovkhozes, the Ministry of Agriculture initiated new hunting and fur-purchasing policies on the smallest kolkhozes in forested parts of the oblast. Siberians had been hunting fur-bearing animals for centuries, but in 1960, for the first time the Ministry of Agriculture took the lead in authorizing and managing the acquisition and sale of furs from the two most profitable creatures of the taiga, squirrels and sables. By creating legal, state-regulated markets for pelts, the state exploited recent population recoveries of these two animals, and it sold their pelts for a significant profit on the international market. By paying a fair price for the furs it purchased, and by regulating the season as well as permits, guns, and other equipment of hunters, the state was able to manage a sustainable, profitable Siberian industry, in stark contrast to the far more ambitious and less successful industrial projects it had initiated in the region.

The fur trade had a much longer history in the area surrounding Lake Baikal than either agriculture or industry. Hunting had been the primary occupation of Siberia's first indigenous groups, the Evenks and the Tofalar. Evenks have continuously inhabited present-day Irkutsk Oblast since the Neolithic period. The Buriats, who arrived later, were originally shepherds, but adopted hunting and trapping in the seventeenth century when the Russian empire began to require annual tribute payments, known as *iasak,* in the form of pelts, from all indigenous groups in Siberia. Throughout the imperial period, the Russian state had limited contact with native groups beyond collecting *iasak* payments, but indigenous trapping and hunting economies were permanently influenced by one social category of Russian European immigrant—the *promyshlennik*, literally a furnisher or middleman, the quint-essential frontier entrepreneur. *Promyshlenniki* were less than scrupulous middlemen who traded with native groups as well as operating independently as hunters and trappers in their own right.

Promyshlenniki introduced two important changes to the fur economy. First, far more often than Russian imperial offi-cials, they used force, extortion, and kidnapping to get what they wanted from local populations. While the Siberian District Of-fice in charge of collecting, appraising, and selling Siberian furs adjusted tribute requirements depending on animal availabil-ity as well as market demands, the less honorable *promyshlen-niki* made no such allowances. The District Office only rarely used extreme force in collecting tribute from native groups, but it gladly purchased all the pelts *promyshlenniki* had to sell, thereby unofficially subsidizing an unregulated and unethical system of trade.[19] Because they made their living off trade, *pro-myshlenniki* introduced European staples such as wheat flour, tobacco, alcohol, and sugar to indigenous groups, which con-tributed to eroding their self-sufficiency.

For the first two centuries that the Russian empire man-aged the Siberian fur trade, the pelts of most animals were relatively plentiful, limited mainly by the Russian state's ability to contact remote indigenous groups and enforce its tribute policies. Until the late nineteenth century it was unimaginable these animals could ever suffer from a population collapse; after all, Siberia was vast, and very thinly populated. During the whole of the imperial period, when Siberia functioned as a colony of Russia, the region's natural resources, not just its furs, seemed abundant and virtually unlimited in supply. Sibe-ria in general hosts three very different ecosystems that stretch like bands from east to west across length of the territory. The northernmost band is tundra, a great semifrozen plain that supports shallow-rooted grasses and lichens. The southern band is another belt of grass, the steppe. Between these bands lies the taiga, the larch and pine forests where Russia's fur-bearing animals live. Irkutsk Oblast in particular seemed destined to produce an unending supply of pelts because so much of the territory was covered with the taiga landscape that sables, ermine, foxes, and squirrels preferred.

The taiga of northern Russia currently contains 22 percent of the world's forested area. For comparison, all of present-day Canada's forests count for less than 7 percent of that area. The majority of Russian forests are taiga, and most of these are located in Siberia, three-quarters of whose territory is forested. Irkutsk Oblast is the most-forested of the Siberian oblasts, with 80 percent of its present-day territory forested, and it has over half of Russia's coniferous forests. In the twentieth century, these forests inspired the construction of lumber mills along the banks of the Angara. During the late Soviet period timber milling and the paper industry became central to the oblast's economy and they remain so to this day, but the fur-bearing animals of the forests also helped Siberia's economic development.

The two most popular pelts emerging from the Siberian taiga were those of the gray squirrel and the sable. European red squirrels were common across the Eurasian continent, but Siberia was home to the gray squirrel, a larger, more heavily furred species that, also due to its remote native habitat, was perceived as more exotic than its European counterpart. Squirrel pelts were plentiful and reasonably affordable to a wide range of consumers. Royalty and wealthy commoners in Europe and the Ottoman Empire wore hats, capes, and coats made from squirrel pelts acquired in Siberia from the fifteenth century onward. The varied but ever-increasing demand from European and Ottoman markets influenced hunting and trade practices in northwestern Russia and Siberia. Demand for sable in particular pushed the early Muscovite empire to establish trading ports along Siberian rivers, effectively colonizing Siberia and lengthening the trade and tribute paths that had already been established with circumpolar hunting peoples in north Russia and the Ural highlands. Russian imperial *iasak* collections made squirrel pelts more available and affordable in the late eighteenth and nineteenth centuries. In contrast, sable, the most expensive and rarest commercially available fur, was typically used only for trim, mainly for the clothing of royalty. Sable was valued and graded principally on its color and glossiness, and although the Russian *iasak* collections meant that more sable pelts were available to European consumers, the fur remained an expensive luxury item.[20]

Until the late nineteenth century market demands and hunters did not threaten the populations of either squirrels or sables. In addition to human predation, the populations of fur-bearing animals in the far north waxed and waned depending on natural pressures such as ground cover and food availability. Gray squirrels were prey for many of the larger forest-dwelling carnivores. Sables, on the other hand, were near the top of the food chain and had few natural predators aside from the occasional Siberian tiger, but human predation was hardly insignificant. By the late nineteenth century the hunting pressures placed on sable made the trade ecologically unsustainable; in various regions between 1870 and 1890, *iasak* requirements were recalibrated so that indigenous groups were allowed to substitute squirrel or beaver pelts in place of sable, a sign that these animals had become more scarce across the Siberian taiga.[21] Sable continued to vanish due to overhunting; they disappeared entirely from forests west of the Urals where they had once been plentiful. In the late nineteenth and early twentieth centuries, the tribute system imperial Russia imposed on native groups across northern Russia evolved into a system of monetary taxation.

Hunting the fur-bearing creatures of Siberia required a high level of skill as well as no small investment in equipment. Hunting sable and squirrel were complementary activities, and most hunters registered as "universal" hunters, meaning they went after any quarry, including foxes, martens, beavers, lynxes, and bears. Well into the mid-twentieth century in Irkutsk, hunters went about trapping, snaring, and shooting their game in ways that closely resembled practices from the seventeenth century and earlier.[22]

Throughout the eighteenth and nineteenth centuries, European immigrants to Siberia who adopted a lifestyle of hunting and trapping instead of farming or trading identified with a new cultural identity, the Sibiriak or Russian Siberian. Such a social category had never before existed, but the model proved durable, and the term is used proudly by European Siberians today. In behavior and appearance, first-generation Sibiriaks were superficially similar to the native groups whose property and livelihood they imitated and at times usurped through extractive trade relations. Rugged and self-reliant, the Russian Sibiriak also resembled the stereotypical mountain man of the American West or the voyageur of Canada, and like these other iconic characters, the Sibiriak became a symbol of nationalist pride, an authentic and legendary member of the pageant of Russian history. Sometimes Sibiriaks operated as *promyshlenniki*, but most of them also hunted and fished. Over time, many of them acquired deeds from the imperial government for property in the arable valleys of Siberia, which meant they also became farmers. While many of the original immigrants to Siberia moved east without spouses or children, intending to make or increase their fortunes before returning to European Russia, a significant number of men permanently settled in Irkutsk and other eastern districts, often settling down with native women and raising mixed-race families.[23]

In the Soviet era in Irkutsk Oblast the ancestors of this first generation of Sibiriaks as well as more recent European immigrants all relied on the fur trade. In postwar Siberia these men—and they were almost exclusively men—still hunted as they had several centuries earlier, on foot either individually or in groups of three or four.[24] Their equipment included dogs, rifles, thick boots, and jackets. The one difference was that until 1955, all sable hunting and some squirrel hunting were banned in the region, so their activities were illegal. Since 1900, sable populations had been so low that the animals had almost disappeared from the registries created by the state.

The Soviet Union launched a massive program to reintroduce sables across Siberia between 1940 and 1965, although the most intensive work in this area was completed by 1952. Between 1930 and 1970, 8,338 sables were released in the eastern Siberian oblasts of Irkutsk, Buriat, Chitinsh, and Yakutsk, the vast majority between 1947 and 1953, in small groups of twenty to thirty individuals. By 1955 studies by Soviet geographers and biologists determined that the animal had recovered and was plentiful across most of Irkutsk Oblast. Since the state had recently invested significant amounts of money in sable reintroduction programs, this was not unbiased research. Nevertheless, the scientific population surveys were used as the basis for reestablishing sable hunting quotas in the oblast. One population biologist, Lazarev, who worked for the Ministry of

Agriculture, wrote in 1955 that "at the present time sables have settled in almost all hunting districts. . . . The greatest density of sables has been observed in the districts on the right bank of the Lena River, and also in the northern hunting district collectives." This discovery happily came just before the start of the Virgin Lands Campaign debacle, and in 1960, when the Ministry of Agriculture wanted to augment the work and incomes of Siberian collective farmers, sable hunting was an obvious choice.[25]

The mixed-use kolkhozes the Ministry of Agriculture started in 1961 were not the first ventures the Soviet state made into managing hunting and other nonagrarian occupations in Siberia, but they were the first successful ones. Earlier, the state had established reindeer herding, sheep herding, and fishing kolkhozes for indigenous groups during the 1920s and 1930s. These were essentially enterprises in repression, managed not by the Ministry of Agriculture, but another entity, the infamous Committee of the North, that treated animal hunting and herding as exclusively economic activities for indigenous groups who had traditionally engaged in these occupations as a way of life. The simplistic attitude adopted by the Committee of the North created resentment and gross inefficiencies. Native reindeer hunters and herders sold carcasses to one state authority and then bought back reindeer hides and meat from a different authority. The Committee of the North ignored ceremonial and seasonal patterns of reindeer hunting, and this significantly weakened the relationship reindeer hunters had traditionally maintained with their herds.[26]

The Committee of the North oversimplified the economies of far northern groups, lumping them into exclusive categories that were not necessarily a good fit with their traditional mixed economies. Collective reindeer ranches were created for indigenous groups that herded the animals for part of the year, but had also previously hunted and fished. Nomadic pastoralists who traded with settled groups and sold their own processed leather goods became members of livestock kolkhozes where sheep ranching was their only task. By simplifying these economies of production, the state also impoverished the groups it forced onto these farms.[27]

Poverty was not just in terms of money. The jobs available to indigenous collective farmers were universally low-paying, but more crucially, they excluded workers from the opportunities for social mobility that allowed many European Soviet citizens of peasant origin to obtain power and prestige in Soviet society during the same period. In the thirty years of Soviet governance indigenous hunting and herding collectives were left unimproved and underdeveloped. Small kolkhozes in Irkutsk Oblast never combined to form larger, better-funded state farms. Before 1956, indigenous kolkhozniks who were found hunting were prosecuted by the state and typically fined, and indigenous people who were not assigned to a reindeer herding collective needed to become licensed hunting "specialists" before they were given permission to kill, regardless of the pre-Soviet economic activities of their ethnic group.[28]

In spite of its stringent hunting laws and policies, during the 1940s and early 1950s the Soviet state was so short on all materials—especially those that might be used for winter clothing or shoes or that could be sold for American dollars—that it advertised in local Irkutsk newspapers, urging city residents to bring all kinds of skins (mentioning rabbit, dog, and seal specifically) to regional slaughterhouses to exchange for cash or credit. The same ad addressed hunters separately: "Hunters! Bring in fur-bearing animals: squirrels, muskrats, weasels, ermine, fox, and sable for special exchange rates!"[29] The special rates advertised in the early 1950s were not particularly impressive; they often amounted to little more than a few rubles and credit at local stores.

It is likely but undocumented that the Soviet state unofficially permitted and even encouraged sable hunting in the late 1950s because of strong demand for pelts from Korea and China until 1960. Officially, sable hunting was banned in the Soviet Union until 1955, although the state recorded purchasing 50,000 pelts from citizens between 1946 and 1955. Sable pelts were far more commonly purchased after hunting became legal again in 1955. Between 1956 and 1965 over 150,000 pelts were purchased, a number that held steady until the late 1970s when sables were successfully domesticated on fur ranches across eastern Siberia.[30]

This new market in sable skins was directly inspired by American Cold War legislation that banned importing luxury goods from the Soviet Union that could be produced domestically. Since squirrels, foxes, and ermine were native to North America, the United States stopped purchasing their pelts from the Soviet Union. These animals were easier for Soviet hunting farms to catch or ranch, process, and market than sable, but after the American ban they were worth less because the pelts could not feed into the large American fur market. For domestic production, Soviet furriers preferred more modest skins that everyday Soviet consumers could afford; nutria was a popular option, and dog fur was not unheard of. While some sable pelts sold for thousands of dollars at the biannual Leningrad fur auction, the fur varied greatly in color and quality, and from year to year it was difficult to predict which kind of pelt would prove the most fashionable and therefore command the highest price. Squirrel, fox, and ermine were more reliable and less varied products, and before the U.S. ban they had been safer products to auction. The ban changed the rules of the game; demand for squirrel, fox, and ermine pelts fell, and demand for sable spiked. Fur exports were an important part of the Soviet Union's economy, and this Cold War–era legislation had serious economic implications. In 1950, Soviet exports to the United States were estimated at 50 million USD, with furs representing 60 percent of the total. Between 1946 and 1955 the state fur industry dressed 400,000 ermine skins, but between 1956 and 1965, the number was only 250,000, with fox and squirrel sales falling correspondingly. In fact, the only animal besides sable whose sales grew after 1950 was the nutria.[31]

Thus northern fur-bearing animals were affected in the 1950s and 1960s by a Cold War ecology in which the lives of Eurasian ermines were spared, and the previously protected sable became the prey of choice for Soviet hunters. The state had the power to select the hunters who were allowed to pursue sables, and in Irkutsk Oblast in the late 1950s, they chose collective farm members who lived and worked in some of the most marginal environments in the Soviet Union on some of the smallest and least profitable farms in the territory. In contrast to the development projects that Irkutsk Oblast hosted between 1955 and 1965, the project to appoint a cadre of hunters within the oblast who would furnish sable and squirrel skins to the state was modest; only a few dozen households were ever granted hunting permits or provided with rifles, bullets, and other hunting equipment. Yet it is striking that this project succeeded admirably, rather than yielding mixed or uncertain results or falling far short of ambitious, abstract goals the state set for it, as had so many projects led by the Ministry of Agriculture.

During the 1950s, the Soviet state had several different and conflicting hunting policies, which stemmed from anxieties about furnishing firearms to untrustworthy citizens. While owning guns was not universally banned, the rules regulating arms ownership were strict. Theoretically, gun ownership was tightly controlled by the state. In practice, many Second World War veterans had simply kept their rifles and service revolvers after the war without bothering to register them or obtain the proper permits to store or use them. The question was especially muddled in Siberia, where members of many hunting and herding kolkhozes were allowed to own guns and administrative overview of ownership and use was lax. In general, native Siberian hunters and many Siberian farmers treated the ownership of a gun as a right while the state treated it as a privilege, and each had tried to ignore the conflicting attitude of the other party.

In 1956 the Ministry of State Farms ceded this question to the Ministry of the Interior, at least as far as it concerned Irkutsk hunters, and the latter was charged with creating a universal arms policy for eastern and northern hunting and herding kolkhozes. The ministry needed to decide "the question of a simplified procedure for receiving permission for the acquisition and retention of guns and cartridges for hunting and reindeer production. . . ."[32] One idea was to lease the guns out to hunters for the relevant part of the year. Another suggestion was to allow gun ownership, but to keep close tabs on ammunition. A third proposed solution was to ban all hunting unless the Soviet state placed specific orders for animals. These discussions mainly concerned hunters and herders on collectives in Siberia and the far north of Russia. Rifle owning was banned on agricultural kolkhozes and this matter was not up for debate.

As part of the research done to determine what might be appropriate firearm legislation, the Ministry of the Interior made a list of northern and eastern oblasts in which residents possessed unregistered rifles. It is not clear how the ministry obtained this information, but the list corresponds closely with the districts of Siberia in which indigenous kolkhozes were located. Bans on hunting were directed at kolkhozes where hunting was part of a way of life, not simply a pastime or a way of providing extra protein. For example, in 1957 on a kolkhoz near Yakutsk, northeast of Irkutsk Oblast, members were forbidden to kill polar bears or walruses, "except in times of acute need" when single walruses could be harvested and collectively portioned out. Visiting scientific expeditions could kill reindeer and walruses to feed their sled dogs, but native Chukhot and Koriak sled dogs (let alone their owners) received no such food allowance, and were expected to make do with state-issued food supplies.[33]

The alternative to native hunters was European hunters, and the Soviet state, like the Russian empire before it, awarded hunting permits to and purchased pelts from Sibiriaks rather than indigenous hunters. These were rural men who also worked as kolkhozniks. In 1958 and 1959 the Ministry of State Farms approved a hunting pilot project at two failing European collective farms, Bolshevik and Taezhnik in the northern part of Irkutsk Oblast. In the Bolshevik project proposal it was estimated that hunter-kolkhozniks could spend one-sixth of their workdays engaged in hunting activities without harming the agricultural economy of the collective farm.[34] Almost all of these hunting days would be taken during the late fall and winter, a relatively dormant season on the farm, and preparing to hunt would occupy the vast majority of the time budget. Hunters needed to construct and repair small shacks across the wilderness for use as base camps, to train hunting dogs, travel to remote hunting sites, and repair guns. Of these activities, building and maintaining the temporary shelters was expected to take up the most time.

From the onset of these projects it was obvious that the Ministry of Agriculture incorporated many more traditional elements into the plan for the hunting cooperatives than was typical for these kinds of projects. On hunting trips, hunters would work in groups of three or four men, usually neighbors, friends, or family members. A group of hunters could divide the labor of a trip and stay out much longer than a single hunter on his own. This went against the general trend of the 1940s and 1950s to remove this personal aspect from working groups. The Ministry of Agriculture and Nikita Khrushchev in particular felt that small-group work encouraged graft and cronyism, and it was officially discouraged in most sectors of industry and agriculture. In another nod to tradition rather than modern ideals, the most experienced hunters were allowed an opportunity to go into the woods early in the season and alone, often coming back with more animals from this first, solo trip than less experienced groups later on. One hunter named Berezovskii, the best in his kolkhoz, brought back 5 sables and 104 red squirrels after six days of hunting in 1959. In that year one sable skin sold for an average price of 488 rubles (with a range of 450–500 rubles), and squirrel skins typically sold for 8 rubles each. At the official exchange rate of 2 rubles to the dollar, this would be equal to $244 per sable and $4 per squirrel. This rate was skewed by Soviet Cold War policies of economic isolation; a more reasonable exchange would be about five to one, making each sable worth approximately $100 and each squirrel worth about $1.60, still an impressive amount for a collective farm worker. The Popamarychuk family on the Taezhnik kolkhoz earned 2,482 rubles from kolkhoz work (not including their in-kind grain payment) and 3,590 rubles from the single two-week trip in 1958 in which they caught 6 sables and 140 squirrels.[35] These very successful hunters were probably not typical. A more realistic review of what sable hunting added to the household economies of Irkutsk kolkhoz members might come from the experience of the family Zchev, also from Taezhnik, whose kolkhoz work netted 3,189 rubles (plus grain) and whose hunting brought in 1 sable and 140 squirrels for a total of 1,380 rubles. Even this more modest take increased the Zchev household income by over one-third. A single sable made an enormous difference to every household that was able to harvest one. Turning failing agricultural kolkhozes into money-making, mixed-use farming and hunting kolkhozes was a limited success. Hunters earned a little extra money, the state earned lots of extra money, and while native groups lost out, the Sibiriak identity, forged in the seventeenth century, was revived.

One of the best indicators of this revival was the sable hunter's dog, the loyal laika breed of eastern Siberia. These dogs were one of the most important tools of a Siberian hunter; dogs were indispensable for tracking and treeing sables. Just as sable hunters tended to hunt in small teams, a pack of half a dozen laikas would hunt for sable jointly.[36] The word *laika* means "barker" in Russian, and as the name suggests the dogs are barking hunters, isolating their prey in the branches of a tree and guarding it from below while vocalizing so that their masters can locate them. The sable would then be finished off, ideally with a single bullet through the head so as not to mar the rest of the pelt. Laikas were valued for their speed, shrill barking, and their ability to work in teams.

The new market in sable fur meant that the Soviet state was inspired to revalue its laika breed of dogs. It is uncertain what native Siberian group first used the laika as a working dog, but when Europeans arrived in Siberia, they noted laika-like dogs living among most of the indigenous groups they encountered. In addition to hunting, laika dogs were trained to perform a variety of tasks including herding reindeer, guarding camps, and pulling sleds. Laikas had not been thought of as an official, deliberately created breed during their long history in Siberia, but this status changed over the course of the twentieth century. Two other dog breeds with origins in the Russian arctic, samoyeds and Siberian huskies, gained international attention because of their role in delivering life-saving diphtheria medicine by sled to remote regions of Alaska in 1925. Perhaps it was the rising fame of these breeds that inspired the Soviet Institute for Hunting Industry to publicly classify the laika as an official breed in 1947. In 1950, the institute also announced that laikas were in danger of dying out; when the population of sables crashed in the early twentieth century, so did the that of laikas. In resurrecting the laika by instituting a breeding program, the Institute for Hunting Industry purposely linked the breed with the Sibiriak lifestyle. Laika-like dogs had also existed for centuries in Finland and Korea, but when the Institute for Hunting Industry created a breeding farm for laikas in Irkutsk, it only accepted dogs from three Siberian oblasts—Irkutsk, Yakutsk, and Novosibirsk—and emphasized the long-standing jobs of hunting and guarding that laikas had performed on the Siberian frontier.[37]

The rising fortunes of the laika paralleled the rise of sable hunters across Irkutsk Oblast. Much like the laika breed, twentieth-century Sibiriaks had been in danger of becoming irrelevant and losing their elevated historic status as early European inhabitants of Siberia. Their economic roles as hunters and occasional traders were threatened by collectivization and industrialization in the postwar period. Multiskilled settlers who had once combined three or more ways of life in order to survive and thrive in the taiga of Irkutsk had been reduced to work at a single, unsuccessful occupation—that of farming. The revived interest in sable hunting coincided with a new movement toward redefining Soviet nationalism away from the multiethnic vision it had embraced before and immediately after the Second World War and toward a specifically Russian version of empire. Resurrecting laika dogs and Sibiriak hunters was part of this trend.

The Soviet preference for European over native hunters in Irkutsk Oblast exposes the shifting goals of the fur industry. In some ways, little had changed from the early days of the imperial tribute system. The state still set annual quotas and fixed prices for furs, Moscow still profited fantastically from the labor of its most distant subjects, and sables were still the most precious luxury item the Soviet Union exported. Likewise, the techniques of hunting had not changed much since the early modern period when Evenks, Nenets, Chukchi, and other indigenous groups pursued the same quarry with dogs, snares, and simple rifles. The work, the landscape, and the product were identical to those of the time when sable hunting had been a traditional subsistence activity of the indigenous inhabitants of Siberia. However, beginning with the first hunting kolkhozes in 1958 and 1959, the Soviet Ministry of Agriculture was the managing authority of these enterprises, and it is significant that these smaller ventures succeeded while the much larger projects failed.

In choosing to provide new economic opportunities to marginal collective farmers the Soviet Union reinforced its policy of settling Siberia and making the territory modern. The majority of hunters registered as sable procurers in 1960 had Russian or Ukrainian last names and were paid for their efforts in sums that far exceeded the wages of other work available to them in the oblast. By investing in European hunters, the Soviet Union practiced selective memory, a tried and true method of state building. It chose to remember the Sibiriak, the part of its past that had established European fur hunting as a traditional practice that enriched individuals, communities, and the state's coffers, and it chose to forget the even longer-standing traditions of tribute and the Russian state's original reliance on the local knowledge and skills of indigenous communities.

The other interventions the state made into Irkutsk Oblast in this period were less successful than the hunting cooperatives. Because the oblast was so cold and so environmentally inhospitable to dense human settlement, progress and prosperity were always expected but never arrived in spite of the state's concerted efforts to triumph over environmental limitations. Although the Bratsk dam was completed ahead of schedule and the state farms of Irkutsk remained resolutely in place, churning out inadequate supplies of meat, milk, and fresh vegetables at tremendous cost to the state, neither industry proved to be a bellwether for further settlement and development.

Like the Russian empire before it and the Russian Federation after it, the Soviet Union overinvested in Siberia in order to achieve a series of near-sighted goals that yielded short-term wealth by sacrificing long-term sustainability. Extracting Siberia's remote wealth and putting it to work for the rest of the Soviet Union was a time-tested but ultimately very expensive form of empire building.[38] The Territorial Production Complexes and the Virgin Lands Campaign were showcase projects designed to prove the country's industrial prowess. The small hunting co-ops the state established as an afterthought as a way to boost salaries and ensure a supply of globally saleable sable pelts were far more basic efforts. These modest campaigns succeeded and thrived while the larger campaigns failed. The internal colonization in Siberia was most successful when it relied on low-tech interventions that made use of significant local manpower and ignored grandiose goals such as nation building. As with the watchful eye of the postwar quarantine stations, the state's attempts to influence and control the rural population of the Soviet Union were most effective when they blended older, established forms of surveillance and work patterns with a modest level of centralized bureaucracy.

Epilogue

In October 1963, the Soviet Union purchased wheat from the United States for the first time since the Second World War. After the deal was announced in the United States, rumors circulated that the Soviet Union planned to use the grain to make explosives, or to convert it into vodka. For Americans during the Cold War era it was easy to imagine the Soviet Union had malevolent or wasteful ulterior motives in brokering a deal like this, and hard to believe that the purchase was driven by pragmatism and food scarcity.[1] President Kennedy's office defended the sale as a savvy capitalist move, saying it "would demonstrate to the world the superiority of free agriculture and free enterprise over the communist collectivist system."[2] For the Soviet Union the wheat purchase was a watershed moment because it marked the beginning of the end of the country's struggle for agricultural self-sufficiency.

The wheat deal signaled a crucial shift in the Soviet Union's agricultural goals. The purchase had little to do with the tensions of the Cold War, instead it was a domestic policy decision for the Soviet Union that would maintain low food prices and improve the quality of animal feed. Some of the grain purchased from abroad was processed into bread products for human consumption, but most of it was fed to livestock. Agricultural planners had tried to rely exclusively on corn silage to feed their cattle and pigs for nearly a decade, ever since Khrushchev had made corn production central to his Virgin Lands Campaign. Like so many other ambitious goals of Soviet agriculture, this had not worked as planned. Corn silage was low in protein, and this affected the productivity of livestock: meat animals gained weight more slowly and developed less muscle mass, and milk cows were slower to start lactating and produced less milk, than their counterparts on a high-protein diet. Even with the personalized care that Soviet milkmaids and swineherds lavished on their charges throughout the Khrushchev era, productivity lagged. In 1960 the cows of the Soviet Union were producing an average of little more than one gallon of milk each day.[3] To put this in perspective, an American Holstein cow during this era produced, on average, between four and eight gallons of milk per day. Through more than a decade of intensive effort, the Soviet Ministry of Agriculture had succeeded in increasing animal stocks around the country, but since there was little to feed them, they could not reach their full potential.

In 1963, facing criticism from his advisers but in solidarity with the malnourished cows of his country, Nikita Khrushchev bought U.S. grain to help make up for anticipated crop losses from droughts and other environmental disasters that had occurred that year. The purchase was domestically unpopular and contributed to Khrushchev's removal from office just one year later.[4] Purchasing wheat from the United States was a significant admission by the Soviet Union that the previous decade of frenetic agricultural reform had not been a complete success. Rather than scale back animal agriculture, Khrushchev decided to change course, shifting away from a policy of agricultural self-sufficiency and toward a more interdependent relationship with world markets, a role that the Soviet Union had shunned but from which it had also been purposely excluded by its Cold War rivals since the end of the Second World War.

Minding the Gap

What lessons can be learned from the thirty-three years the Soviet Union spent striving for agricultural self-sufficiency? The Soviet case yields a few important insights. First, there were often gaps between planned accomplishments and reality, but this did not necessarily mean that unexpected or less than perfect outcomes were always failures. Soviet agricultural reforms set out to prove to the world that socialist agricultural science and management offered a feasible alternative to the exploitative successes of capitalist agribusiness. In the five examples discussed here, the state's initial ambitions were larger than the results that were ultimately realized, but in none of these cases can the Soviet Union's reforms and interventions be judged to be a complete failure.

Soviet officials often pretended that the ambitious goals the state had set for agriculture were being met or exceeded when in fact the opposite was true; industrial progress on farms was modest and occurred in fits and starts. Because agriculture depended on environmental factors that were beyond the control of the state, it was far more vulnerable than other sectors of the economy to small mishaps and miscalculations; a single drought or early frost could prove devastating to harvest levels or animal reproduction rates. Nevertheless, it was always in the best interests of the Soviet government to pretend that its plans and reforms were going well. Particularly when it came to agriculture, the communist system specialized in triumphal official descriptions that left little room for self-reflection or critical analysis. When American scientists and economists discovered the Soviet Union lying about its agrarian successes, they invariably read too much into these lies, accepting them as proof that the struggling system was on the brink of collapse. A second insight that emerges from these eventful three decades relates to the ulterior motives of agricultural reform. Agricultural systems produced more than food for the Soviet state. Creating and managing collective farms reinforced Soviet power in the countryside and provided a public forum in which the state could trumpet its successes. The Soviet state focused on projects that established stronger scientific and technical bureaucracies in rural areas. The Soviet Union fell short of many of its stated goals to produce more food of better quality, but every agricultural reform was also an opportunity for the state to get a better handle on the vast and poorly governed Soviet countryside. Over the thirty-three years chronicled in this book, the Ministry of Agriculture discovered that agrarian reform held great potential for a state in search of greater political legitimacy and environmental control.

Finally, farming in the Soviet Union was an environmental activity as much as it was cultural or political. This fact has always been true, but Soviet authorities went through different stages of attending to this truth. Early in the history of the Soviet Union, leaders ignored the environment entirely. Later, they insisted they could harness nature and completely control it. Both attitudes resulted in failed policies. The case studies in this book illustrate the evolution of this attitude over time. In the 1930s, when George Heikens and Guy Bush worked on pig farms, the natural environment was not considered an important force that shaped the success of farm modernization. Although Bush and Heikens pushed the government to take into account environmental factors like cold weather and the spread of diseases, they did not frame these as environmental issues. Americans and Soviets both interpreted the successes and failures of these farms as exclusively human in origin.

Immediately after the Second World War, the Soviet Union was more conscious of the power the environment held in safeguarding or stymieing the success of agriculture. In fact, control of nature was a crucial element of many postwar agricultural projects. From waging a "war" against invasive species on the Polish Ukrainian border to deciding by site visit and committee vote that the Lebedinskii cow should be recognized as a new, purebred animal, postwar agricultural management almost always included plans for controlling or efficiently managing natural resources and natural phenomena.

Two Universal Lessons

The Soviet experience holds two universal lessons about the legacy of agricultural modernization that transcend the topics of environmental and agricultural history. The first is the surprising and overlooked utility that the ideologically based science of Lysenkoism possessed for Soviet authorities. The second lesson is that the very public hubris of Soviet modernization schemes was precisely what allowed their flaws and shortcomings to become so starkly apparent. The visible and public nature of many of the Soviet state's missteps in the countryside have become guideposts and warnings for people who want to learn from the past.

Lysenkoism was the Soviet Union's ideological and flawed alternative to genetics, and Lysenko's theories had long-lasting repercussions for plant and animal breeding programs. As a science of animal management, specifically as a management program designed to mitigate the harsh effects of the Soviet climate, Lysenkoism proved to be an effective tool during a critical period of rebuilding and reorganizing. During Khrushchev's years in power, Lysenkoism became an important system through which the state tried to control the Soviet environment. Lysenkoism provided a basic but effectual approach to livestock improvement that allowed the Soviet Union to play to its strengths in animal agriculture: an abundant supply of human labor, and a diminished but environmentally well-adapted population of mixed-breed animals. By ignoring principles of inheritance and selective breeding that the Western science of genetics accepted as laws, the Soviet Union was able to maximize many of its postwar animal resources in spite of significant and long-standing handicaps rooted in both nature and culture.

This was not some sleight of hand; Soviet animals at mid-century were in need of the most basic interventions in order to survive and reproduce. Better food, individualized care, and better living environments helped boost survival rates for animals in the postwar era. It is impressive that the Soviet Union's Ministry of Agriculture managed to mask these improvements as a rigorous scientific program. The power of scientific theories that are useful, intuitive, and easy for the general population to understand should not be underestimated. The faulty science of Lysenkoism did not thrive in the Soviet Union for more than a quarter-century simply because the country had an authoritarian government and its population was easily duped. Lysenkoism thrived because it served an important and necessary function for both the government and the citizens of the Soviet Union It offered a clear and intuitive path toward agricultural improvement.

The unusually visible flaws in Soviet agriculture have rapidly become lessons for the rest of the world in what can go wrong with agricultural modernization. The dissolution of the

Soviet Union and the end of the Cold War helped to illuminate the dysfunctions of the Soviet countryside, but well before these events, the Soviet state's outsized ambitions for its farms opened agriculture to a level of scrutiny that made it impossible for the state to hide the many shortcomings and missteps of its rural reforms. Recently, this has not been the case in the Western countries. Precisely because agricultural progress in the West takes place primarily through a dispersed, private network of capital and power, it is much harder to account for the complex array of influential authorities and financial structures that undergird contemporary agrarian failures. Soviet mistakes were obvious, but this does not mean that capitalist democracies have not also made equally grave errors in how they have chosen to intensify and scale up their agricultural systems; there are simply more layers of concealment obfuscating the shortcomings and pitfalls of agriculture in these systems.

Learning from the Soviets

Based on the case studies featured here, what can this visible, flawed Soviet experience of agricultural industrialization teach the rest of the world about how to effect change in rural areas? Is it possible that the Soviet Union's agrarian reforms could serve as a model, rather than simply as a warning? The Soviet Union should be given credit for the things it did right in its quest to improve its farms. Soviet modernization did not occur in a vacuum; Soviet experts looked to United States and Western Europe for models of progressive agriculture. In its technological lag, the Soviet state had an advantage because agriculture authorities were able to cherry-pick promising technologies, selecting only those that had succeeded elsewhere and were truly relevant to the situations at hand.[5] The Soviet Union's agricultural planners proved themselves adept at investing money and effort in technologies they believed would grant the best return. Regardless of the ideological gulf that separated the Soviet Union from its capitalist counterparts in the West, the Soviet Union participated in many exchanges of technology and science, almost always to the benefit of the Soviets, who were shrewd collectors of such expertise.

The Ministry of Agriculture and other national-level ministries also learned to tolerate a certain level of chaos and failure at the local level. For much of the 1930s, historians have persuasively argued that the eye of the Soviet state was often focused intently on Soviet citizens and their behavior. This was especially true on the newly formed collective farms. After the Second World War, the Soviet government retreated from these tactics. This was initially because state resources were thin immediately after the war, but throughout Stalin's tenure, and increasingly during Khrushchev's period of leadership, rural micromanaging was replaced by less extensive targeted surveillance.

This new tactic, in which the state chose its agrarian battles, blended an air of false bravado with an ability to respond to local circumstances which became a hallmark, not just of the reform-minded Khrushchev years but also of the Stalinism that preceded them. This spirit of compromise had very little to do with the inherent generosity or flexible nature of the Ministry of Agriculture. Instead, as time passed and leaders within the Ministry of Agriculture gained experience and perspective, they learned that droughts, cold snaps, heat waves, muddy roads, contagious diseases, and soil erosion could confound even the most ambitious reforms. It became essential that plans could be modified and adapted to local conditions. It was the stubborn and intractable nature of the Soviet environment that ultimately forced the Soviet agricultural system to come to terms with itself.

Archives and Contemporary Periodicals

Archives in Russia

Gosudarstvennyi arkhiv Irkutskoi oblasti, Irkutsk
  Fond R-147
Gosudarstvennyi arkhiv noveishei istorii Irkutskoi oblasti, Irkutsk
  Fond 127
Gosudarstvennyi arkhiv rossiiskoi federatsii, Moscow
  Fond 616
  Fond R-7021
Rossiiskii gosudarstvennyi arkhiv ekonomiki, Moscow
  Fond 8390, VASKhNIL (All-Union Agricultural Institute)
  Fond 8295, MinMiasProm (Ministry of Meat Production)
  Fond 8297, MinMiasProm (Ministry of Meat Production)
  Fond 7468, MinSelKhoz (Ministry of Agriculture)
  Fond 7803, MinSovKhoz (Ministry of Sovkhoz Production)

Archives in Ukraine

Derzhavnii arkhiv Kharkivs'koi Oblasti, Kharkov
  Fond R-6184
  Fond R-4447
Tsentral'nii derzhavnii arkhiv vyshykh organiv vladi, Kiev
  Fond R-27

Archives in the United States

Iowa State University Library Special Collections, Ames, IA
    George A. Heikens Papers (1902–1976)
    J. Marion Steddom Papers
    Lauren Soth Papers (1945–1976)
John F. Kennedy Presidential Library, Boston
National Archives and Records Administration, College Park, MD
    RG 59, General Records of the Department of State
    RG 166, Foreign Agricultural Service Narrative Reports
    Smolensk Archive Microfilm Series
Private collection of Guy L. Bush Jr., East Lansing, MI
    Letters of Guy L. Bush to Louise G. Bush, 1930–31
Widener Library, Harvard University, Cambridge, MA
    Harvard Project on the Soviet Social System

Newspapers and Journals

Izvestiia, Moscow
Kholodil'naia tekhnika, Moscow
Kolkhoznoe proizvodstvo, Moscow
Kolkhoznoe selo, Kiev
Molochnaia promyshlennost', Moscow
Pishchevaia promyshlennost' (Molochnaia), Moscow
Pravda, Moscow
Sotsialisticheskoe zemliadelie, Moscow
Voprosy pitaniia, Moscow
Vostochno-Sibirskaia Pravda, Irkutsk

Notes

Citations to Russian and Ukrainian archival documents are by *fond, delo, opis'*, and *list* (f., d., op., l.). Archive names are abbreviated as follows (for full details, see the List of Archives and Contemporary Periodicals).

Bush Collection	Private collection of Guy L. Bush Jr., East Lansing, MI
DAKhO	Derzhavnii arkhiv Kharkivs'koi Oblasti, Kharkov
GAIO	Gosudarstvennyi arkhiv Irkutskoi oblasti, Irkutsk
Harvard Project	Harvard Project on the Soviet Social System, Widener Library, Harvard University
Heikens Papers	George A. Heikens Papers (1902–1976), Iowa State University Special Collections
NARA	National Archives and Records Administration, College Park, Md.
RGAE	Rossiiskii gosudarstvennyi arkhiv ekonomiki, Moscow
Smolensk Archive	Smolensk Archive Microfilm Series, NARA
TsDAVO	Tsentral'nii derzhavnii arkhiv vyshykh organiv vladi, Kiev

Introduction

1. On literacy during the late imperial period, see Ben Eklof, "Russian Literacy Campaigns 1861–1939," in *National Literacy Campaigns and Movements: Historical and Comparative Perspectives,* ed. Robert F. Arnove and Harvey J. Graff (New Brunswick, NJ: Transaction, 2008), 128–29. On the rural-urban divide see Nicholas Spulber, *Russia's Economic Transitions: From Late*

Tsarism to the New Millennium (Cambridge: Cambridge University Press, 2003), 140. On animal populations, see Orlando Figes, *Peasant Russia, Civil War: The Volga Countryside in Revolution, 1917–1921* (New Haven, CT: Phoenix, 2001).

2. V. I. Lenin, *Collected Works,* vol. 9, trans. Abraham Fineberg and Julius Katzer (Moscow: Progress, 1972), 232–33.

3. Notable works by dissidents and refugees include Viktor P. Danilov et al., *Tragediia sovetskoi derevni: Kollektivizatsiia i raskulachivanie. Dokumenty i materialy, 1927–1939,* 5 vols. (Moscow: Rosspen, 1999); Miron Dolot, *Execution by Hunger: The Hidden Holocaust* (New York: Norton, 1987); Zhores Medvedev, *Soviet Agriculture* (New York: Norton, 1987); Moshe Lewin, *Russian Peasants and Soviet Power: A Study of Collectivization,* trans. Irene Nove with John Biggart (London: George Allen & Unwin, 1968); and Maurice Hindus, *Red Bread: Collectivization in a Russian Village* (1931; Bloomington: Indiana University Press, 1988). Works by Cold War–era scholars include Robert Conquest, *Harvest of Sorrow: Collectivization and the Terror-Famine* (Oxford: Oxford University Press, 1987); Martin McCauley, *Khrushchev and the Development of Soviet Agriculture: The Virgin Land Programme, 1953–1964* (New York: Holmes & Meier, 1976); Robert F. Miller, *One Hundred Thousand Tractors: The MTS and the Development of Controls in Soviet Agriculture* (Cambridge, MA: Harvard University Press, 1970).

4. Tania Murray Li, *The Will to Improve: Governmentality, Development, and the Practice of Politics* (Durham, NC: Duke University Press, 2007).

5. Stephen Brain, *Song of the Forest: Russian Forestry and Stalin's Environmentalism* (Pittsburgh: University of Pittsburgh Press, 2011).

6. Donald Worster, "Transformations of the Earth: Toward an Agro-Ecological Perspective of History," *Journal of American History* 76, no. 4 (March 1990): 1087–1106.

7. Cf. Michaela Pohl, "'The 'Planet of 100 Languages': Ethnic Relations and Soviet Identity in the Virgin Lands," in *Peopling the Russian Periphery: Borderland Colonization in Eurasian History,* ed. Nicholas Breyfogle, Abby Schrader, and Willard Sunderland (London: Routledge, 2007); and Niccolo Piancola, "Famine in the Steppe: The Collectivization of Agriculture and the Kazak Herdsmen, 1928–1934," *Cahiers du monde russe* 45, nos. 1–2 (2004): 137–92.

Chapter 1
Model Farms and Foreign Experts

1. This is similar to James Scott's notion of *metis,* featured in *Seeing Like a State: How Certain Schemes to Improve the Human Condition Have Failed* (New Haven, CT: Yale University Press, 1998).

2. Esther Kingston Mann, *In Search of the True West: Culture, Economics and Problems of Russian Rural Development* (Princeton, NJ: Princeton University Press, 1999).

3. In addition to the two men's letters, Bush also published several articles about his travels: Guy Bush, "Nine Out of Ten Pigs Died: Iowa Farmer Tells of Raising Hogs in Russia," "Where Hired Men Issue Orders," and "What Is Russia's Major Vice?" *Wallaces Farmer*, 1 and 28 November, 26 December 1931.

4. Josef Stalin, *Works*, vol. 14 (London: Red Star, 1978), 153; *Sistematicheskoe sobranie zakonov RSFSR, Ukazov Prezidiuma Verkhovnogo Soveta RSFSR* [1921], Dekrety Sovetskoi Vlasti, vol. 13 (Moscow: Rosspen, 1989), 250–51.

5. My understanding of NEP in the Soviet countryside is based on the works of Viktor Danilov, *Rural Russia under the New Regime* (Bloomington: Indiana University Press, 1988); Susan Gross Solomon, *The Soviet Agrarian Debate: A Controversy in Social Science, 1923–1929* (Boulder, CO: Westview, 1977); James W. Heinzen, *Inventing a Soviet Countryside: State Power and the Transformation of Rural Russia, 1917–1929* (Pittsburgh: University of Pittsburgh Press, 2004); and Moshe Lewin, *Russian Peasants and Soviet Power: A Study of Collectivization* (New York: Norton, 1975).

6. These debates occurred in the context of a larger set of political disagreements that solidified Stalin's leadership and alienated Bukharin, Trotsky, and other prominent Soviet leaders. Stalin did not publicly support mass collectivization or the whirlwind approach until 1928. Cf. Alexander Erlich, *The Soviet Industrialization Debate, 1924–1928* (Cambridge, MA: Harvard University Press, 1960) and Alec Nove, *An Economic History of the U.S.S.R.* (Baltimore: Penguin, 1969), chaps. 5–6. On collectivization debates, see Lynne Viola et al., *The Tragedy of the Soviet Countryside: The War Against the Peasantry, 1927–1930*. vol. 1 (New Haven, CT: Yale University Press, 2005), and Nikolai Alexeevich Ivnikii, *Repressivnaia politika sovetskoi vlasti v derevne, 1928–1933 gg.* (Moscow: Inst. Rossiskoi Istorii RAN, 2000).

7. Lynne Viola, *Peasant Rebels Under Stalin* (Oxford: Oxford University Press, 1999), 71.

8. On collectivization protests, see Tracy McDonald, "A Peasant's Rebellion in Stalin's Russia," in *Contending with Stalinism: Soviet Power and Popular Resistance in the 1930s*, ed. Lynne Viola (Ithaca, NY: Cornell University Press, 2002); Mark Tauger, "Soviet Peasants and Collectivization, 1930–1939: Resistance and Adaptation," in *Rural Adaptation in Russia*, ed. Stephen K. Wegren (New York: Routledge, 2005); and Viola, *Peasant Rebels under Stalin*.

9. Moshe Lewin, "The Disappearance of Planning in the Plan," *Slavic Review* 32, no. 2 (June 1973): 287.

10. "Spetsialistu po svinovodstvu Kheikinsu: Instruktsii," 1930, R/S 21/7/15, Box 1, Folder 2, Heikens Papers.

11. Merle Fainsod, *Smolensk Under Soviet Rule* (Cambridge, MA: Harvard University Press, 1958), 278.

12. Tania Murray Li, *The Will to Improve: Governmentality, Development, and the Practice of Politics* (Durham, NC: Duke University Press, 2007); Lewis Feuer, "American Travelers to the Soviet Union 1917–1932: The Formation of a Component of New Deal Ideology," *American Quarterly* 14 (June 1962): 119–49.

13. See Paul Josephson, *Would Trotsky Wear A Bluetooth? Technological Utopianism Under Socialism* (Baltimore: Johns Hopkins University Press, 2009), 193–232; and Douglas R. Weiner, *Models of Nature: Ecology, Conservation, and Cultural Revolution in Soviet Russia* (Pittsburgh: University of Pittsburgh Press, 2000).

14. Fainsod, *Smolensk Under Soviet Rule*, 20; "Doklad zapiska ot novobytova," Smolensk Archive, 1930, T87, Roll 20, File 169, 2; Christine D. Worobec, *Peasant Russia: Family and Community in the Post-Emancipation Period*, (Evanston: Northern Illinois University Press, 1995), 17–42; Fainsod, *Smolensk Under Soviet Rule*, 22; "Doklad zapiska ot novobytova," 1930, T87, Roll 20, File 169.

15. Heikens to Heikens family, 8 August 1930, R/S 21/7/15, Box 1, Folder 5, Heikens Papers.

16. Bush to Louise G. Bush, 16 September 1930, Bush collection.

17. Bush to Louise G. Bush, 4 March 1931, Bush collection. This description is similar to reports of "Amerikanka," the enclave originally constructed for American engineers at Magnitogorsk. Stephen Kotkin, *Magnetic Mountain: Stalinism as a Civilization* (Berkeley: University of California Press, 1995), 125–26. Kotkin writes that foreign engineers were similar in status to professionals housed on farms before collectivization. See also James Heinzen, "Professional Identity and the Vision of the Modern Soviet Countryside: Local Agricultural Specialists at the End of NEP 1928–1929," *Cahiers du monde russe* 39, no. 1–2 (1998): 9–26.

18. Heikens to Heikens family, 20 August 1930, Box 1, Folder 5, Heikens Papers; Bush to Louise G. Bush, 21 December 1930, Bush collection.

19. Bush to Louise G. Bush, 20 September 1930, Bush collection.

20. Heikens to Heikens family, 4 August 1930, Box 1, Folder 5, Heikens Papers; Heikens to Heikens family, September/October 1930, Box 1, Folder 5, Heikens Papers.

21. 23 August 1930, Box 1, Folder 16, Heikens Papers.

22. Bush to Louise G. Bush, 20 September 1930, Bush collection.

23. Heikens to Heikens family, 20 October 1930, R/S 21/7/15, Box 1, Folder 5, Heikens Papers; Bush to Louise G. Bush, 4 October 1930, Bush collection.

24. Fainsod, *Smolensk Under Soviet Rule,* 270.

25. Ibid.

26. Bush to Louise G. Bush, 4 October 1930, Bush collection; Heikens to Heikens family, September/October 1930, Heikens Papers.

27. Heikens to Heikens family, 8 August 1930, R/S 21/7/15, Box 1, Heikens Papers; Bush to Louise G. Bush, 20 October 1930, Bush collection; Feuer, "American Travelers to the Soviet Union," 120–21.

28. Heikens to Heikens family, 29 August, 7 September 1930, and n.d. (December) 1930, Box 1, Folder 5, Heikens Papers.

29. Bush to Louise G. Bush, 14 December and 24 November 1930, Bush collection; Fainsod, *Smolensk Under Soviet Rule,* 265.

30. Heikens to Soviet Swine Trust, 15 January 1930, R/S 21/7/15 Box 1, Folder 3, Heikens Papers.

31. Heikens to Heikens family, 16 February 1930, R/S 21/7/15 Box 1, Folder 5, Heikens Papers.

32. Bush to Louise G. Bush, 24 November 1930, Private collection of Guy L. Bush Jr.; Heikens to the Swine Trust, 4 April 1930, R/S 21/7/15 Box 1, Folder 5, Heikens Papers.

33. Compare the experience of Samuel Harper in the Soviet Union in 1926 in David Engerman, *Modernization from the Other Shore: American Intellectuals and the Romance of Russian Development* (Cambridge, MA: Harvard University Press, 2003), 129–32.

34. Heikens to Heikens family, 10 April 1930, R/S 21/7/15 Box 1, Folder 5, Heikens Papers; Heikens to the Swine Trust, 4 April 1931, R/S 21/7/15 Box 1, Folder 8, Heikens Papers.

35. Bush to Louise G. Bush, 12 October 1930, Bush collection.

36. Bush to Louise G. Bush, 17 November 1930, Bush collection.

37. Maurice Hindus notes the same cramped barn layout and lack of glass (due to cost) on farms in 1931. Maurice Hindus, *Red Bread: Collectivization in a Russian Village* (Bloomington: Indiana University Press, 1988), 20, 232.

38. Heikens to Heikens family, 10 April 1930, R/S 21/7/15 Box 1, Folder 5, Heikens Papers.

39. Bush to Louise G. Bush, 20 September 1930, Bush collection.

40. Heikens to Heikens family, 4 April 1930, R/S 21/7/15 Box 1, Folder 5, Heikens Papers.

41. Smolensk Archive, T87 Roll 8, location 156, 5–6, and T87 Roll 8, location 57, 2.

42. Bush to Louise G. Bush, 9 August 1931. Bush collection.

43. Dana Dalrymple, "The American Tractor Comes to Soviet Agriculture: The Transfer of a Technology," *Technology and Culture* 5, no. 2 (1964): 206.

Chapter 2
Restoring Control

1. The experiences and failures of this group are detailed in Lynne Viola, *The Best Sons of the Fatherland: Workers in the Vanguard of Soviet Collectivization* (Oxford: Oxford University Press, 1989), 112–20.

2. See Arvid Nelson, *Cold War Ecology: Forests, Farms, and People in the East German Landscape, 1945–1989* (New Haven, CT: Yale University Press, 2005), 29–52.

3. The Soviet Ministry of Agriculture was known as the People's Commissariat of Agriculture, or Narkomzem, prior to 1946. For consistency, I will refer to this institution as the Ministry of Agriculture throughout this chapter and for the remainder of the book.

4. RGAE, f. 7468, op. 12, d. 1609. This report outlines the number and acreage of kolkhozes and sovkhozes in Kharkov Oblast. The number of kolkhozes drops from 1,985 in 1948 to 1,465 in 1950.

5. Harvard Project, Schedule B, vol. 10, Case 15 (interviewer A. D.), 1–2, and vol. 12, Case 30 (interviewer H. D.), 7.

6. Zhores Medvedev, *Soviet Agriculture* (New York: Norton, 1987), 129–30: "O Razukrupnenni Zernovykh Sovkhozov," decree issued by the Council of People's Commissars of the USSR, 22 December 1933.

7. Harvard Project, Schedule B, vol. 16, Participant 20, (interviewer H. D.), Widener Library; Julie Hessler, *A Social History of Soviet Trade: Trade Policy, Retail Practices and Consumption, 1917–1953* (Princeton, NJ: Princeton University Press, 2004), 300.

8. Elena Zubkova, *Russia after the War: Hopes, Illusions, and Disappointments, 1945–1957* (Armonk, NY: M. E. Sharpe, 1998), 36.

9. "The MTSs were rural agencies that supplied collective farms with agricultural machinery and people to run it. They were set up in the late 1920s and early 1930s, when the kolkhozes were too weak and disorganized to manage their own equipment. . . . Politically, the new collective farms, into which so many peasants were dragooned, were unreliable. So the MTS also served as a party (and police) stronghold in the countryside." William

Taubman, *Khrushchev: The Man and his Era* (New York: Norton, 2004), 375.
See also Robert F. Miller, *One Hundred Thousand Tractors: The MTS and the Development of Controls in Soviet Agriculture* (Cambridge, MA: Harvard University Press, 1970).

10. RGAE, f. 7486, op. 12, d. 1414.

11. RGAE, f. 7486, op. 12, d. 1339, ll. 1–5.

12. Viola, *The Best Sons of the Fatherland*, 152–78.

13. Michael Ellman, "The 1947 Soviet Famine and the Entitlement Approach to Famines," *Cambridge Journal of Economics* 24, no. 5 (2000): 605; Ellman, "The 1947 Soviet Famine," 618–19; and Nicholas Ganson, *The Soviet Famine of 1946–47 in Global and Historical Perspective* (New York: Palgrave Macmillan, 2009).

14. Ellman, "The 1947 Soviet Famine," 613–14; "USSR: Crop Estimating 1949–1946," RG 166, Foreign Agricultural Service Narrative Reports 1946–1949, Box 1002, NARA.

15. Joseph Bulik, "Report on a Second Day's Visit in the Kiev Region of the Ukraine," 7 August 1946, RG 166, Foreign Agricultural Service Narrative Reports 1946–1949, Box 1002, NARA.

16. Averill Harriman to Secretary of State, telegram, 4 April 1945, U.S. Department of State, *Foreign Relations of the United States: Diplomatic Papers, 1945. Europe*, 819.

17. Harry S. Brown (of the U.S. Agricultural Service) to President Truman, n.d. (1947), 861.61311/11-47, LM 176, Reel 30, State Department Records, NARA: "According to newspaper reports, Soviet Russia has a substantial surplus of wheat. . . . I respectfully suggest that the United State Government should publicly offer to credit Soviet Russia against her Lend-Lease indebtedness for all wheat she is willing to give to the European countries participating in the Marshall Plan."

18. Ellman, "The 1947 Soviet Famine," 618–19. This statement is also based on research into the famine and its causes, cited above, note 13.

19. Redcliffe N. Salaman, *The History and Social Influence of the Potato*, rev. ed. (Cambridge: Cambridge University Press, 1985), 130, 136.

20. Leon Martel, *Lend-Lease, Loans, and the Coming of the Cold War: A Study of the Implementation of Foreign Policy*, Westview Special Studies in International Relations (Boulder, CO: Westview, 1979).

21. Zhores Medvedev, *Soviet Agriculture* (New York: Norton, 1987), 224.

22. Ibid., 143–44.

23. William Moskoff, *The Bread of Affliction: The Food Supply in the USSR During World War II* (Cambridge: Cambridge University Press, 2002),

154–55; Hessler, "Vodka Production and Sales," in *A Social History of Soviet Trade*, Appendix C, online at http://darkwing.uoregon.edu/~hessler/appen dix/Vodka.htm.

24. My knowledge about the Russian empire's experiences with cholera and disease quarantine comes from Peter Baldwin, *Contagion and the State in Europe, 1830–1930* (Cambridge: Cambridge University Press, 2005); and Charlotte Henze, *Disease, Health Care and Government in Late Imperial Russia: Life and Death on the Volga, 1823–1914* (Taylor & Francis, 2010), 64.

25. Henze, *Disease, Health Care and Government*, 59–60.

26. Stephen Lacey, "Cholera: Calamitous Past, Ominous Future," *Clinical Infectious Diseases* 20, no. 5 (May 1995), 1412.

27. RGAE, f. 7486, op. 12, d. 1339, l. 240; Ol'ga Yu. Elina, *From the Tsar's Gardens to Soviet Fields: A History of Agricultural Experimental Institutions, XVIII c. to the 1920s* (in Russian), 2 vols. (Moscow: Egmont-Russia, 2008), introduction.

28. RGAE, f. 7486, op. 12, d. 1339, l. 14.

29. Ibid.

30. RGAE, f. 7486, op. 12, dd. 1644, 2007, 1414.

31. L. F. Haber, *The Poisonous Cloud: Chemical Warfare in the First World War* (Oxford: Clarendon Press, 1986), 170; and A. M. Prentiss, *Chemicals in War: A Treatise on Chemical Warfare* (New York: McGraw Hill, 1937), 661–66.

32. RGAE, f. 7486, op. 12, d. 1545, ll. 6–7.

33. Benjamin C. Garrett, "The Colorado Potato Beetle Goes to War," *Chemical Weapons Convention Bulletin* no. 33 (September 1996): 2–3; Milton Leitenberg, Raymond A. Zilinskas, and Jens H. Kuhn, *The Soviet Biological Weapons Program: A History* (Cambridge, MA: Harvard University Press, 2012), 407–9.

34. E. P. Ivanschik and S. S. Izhevskii, "*Istoriia koloradskogo zhuka, Leptinotarsa decemlineata*, Dispersal and Its Current Range," in *Koloradskii zhuk, Leptinotarsa decemlineata*, ed. R. S. Ushatinskaia (Moscow: Nauka, 1981), 11–26.

35. RGAE, f. 7486, op. 12, d. 2007, l. 16.

36. My understanding of the workings of the Quarantine Station system derives from the following files: RGAE, f. 7468, op. 12, dd. 1609, 1414, and d. 1339, l. 240.

37. Jeffrey Burds, "The Soviet War against Fifth Columnists: The Case of Chechnya, 1942–1944," *Journal of Contemporary History* 42, no. 2 (April 2007): 268; Edmund Russell, *War and Nature: Fighting Humans and Insects with Chemicals from World War I to Silent Spring* (Cambridge: Cambridge University Press, 2001), 17–36, 95–164.

38. RGAE, f. 7468, op. 12, d. 1609, ll. 5–8.

39. In Kiev, the city with the highest percentage of Party members holding Quarantine Station jobs, 30 of 137 workers were members in 1948, and 102 of 137 were classified as having received a higher education. RGAE, f. 7468, op. 12, d. 1339, l. 7.

40. "Letter from the Embassy to the Department of State, dated September 2, 1948," State Department Decimal Files, LM 176, Reel 30, 861.61/9-248, "1945–1949 Soviet Union Agriculture," NARA.

41. Report of a trip made by Oscar Holden, 30 October 1948, State Department Decimal Files, LM 176, reel 30, 861.61/10–3048, NARA; Letter from Price, 24 September 1948, State Department Decimal Files, LM 176, Reel 30, 861.61/9–2448, NARA.

42. RGAE, f. 7468, op. 22, d. 105, l. 33.

43. RGAE, f. 7468, op. 12, d. 1414.

44. RGAE, f. 7468, op. 12, d. 1414.

45. RGAE, f. 7486, op. 12, d. 1609, l. 9. This was a bacterial blight found in a shipment of tomatoes.

46. *HALT Amikäfer* (Berlin: Amt für Information der Regierung der DDR, 1950), trans. Randall Bytwerk as *STOP Yankee Beetles,*. available online at http://www.calvin.edu/academic/cas/gpa/amikafer.htm.

47. RGAE, f. 7486, op. 12, dd. 1609, l. 8, 1339, ll. 24–27.

48. RGAE, f. 7486, op. 12, d. 1609, l. 1; Cynthia Kaplan, *The Party and Agricultural Crisis Management in the U.S.S.R.* (Ithaca, NY: Cornell University Press, 1987).

Chapter 3
Animal Farms

1. J. A. Newth, "Soviet Agriculture: The Private Sector 1950–1959. Animal Husbandry," *Soviet Studies* 13, no. 4 (1962).

2. DAKhO, f. R-6184, op. 1, d. 302; RGAE, f. 8297, op. 5, d. 3, l. 30.

3. DAKhO, f. 5804, op. 1, d. 51, l. 34. This mimicked the situation in human medicine, where feldshers and public health technicians served in rural areas both before and after the Second World War.

4. Richard Pipes, *Russia under the Old Regime* (London: Weidenfeld & Nicolson, 1974), 6–7.

5. *Five-Year Plan for the Rehabilitation and Development of the National Economy of the USSR, 1946–50*, trans. Nikolai Voznesenskii (London: Soviet News, 1946); "On the Measures to Build Up Agriculture in the Post-War Period," *Pravda*, 28 February 1947.

6. For an overview of this process in an American context, see Richard P. Horwitz, *Hog Ties* (New York: St. Martin's Press, 1998); and *Industrializing Organisms: Introducing Evolutionary History,* ed. Philip Scranton and Susan Schrepfer (New York: Routledge, 2004). A Soviet agricultural delegation to the United States wrote a detailed private report on the topic of American animal industrialization; RGAE, f. 7486, op. 22, d. 89.

7. These numbers had been decided by a decree of the Central Committee in the early 1930s. Alec Nove, "Rural Taxation in the U.S.S.R.," *Soviet Studies* 5, no. 2 (1953): 159–66.

8. For example, see Burlin Hamer, "Trip Report from Moscow to Odessa and Return by Way of Kharkov," May 1951, RG 166, Foreign Agricultural Service Narrative Reports, Box 63, Folder: Agriculture 1954–1950, NARA; *Sotsialisticheskoe zemliadelie,* 22 February 1947, quoted in "Report by an UNRRA Official on Nutrition in the Ukrainian SSR," RG 166, Foreign Agriculture Service Narrative Reports, Box 1000, Folder: Agriculture 1946–1949, NARA.

9. On one breeding farm near Kharkov, 4,959 of 20,360 Ukrainian Grey milk cows were listed as "purebred or (stable) bred animals with no birth records." TsDAVO, f. R-27, op. 18, d. 6671.

10. See various authors, "United States–Russian Farm Delegates Exchange: Correspondence," 16/3/54 Box 11 70–71, Lauren Soth Papers, Iowa State University Special Collections.

11. DAKhO, f. 5804, op. 1, d. 5, l. 38. Railroad workers had the right to cultivate right-of-ways. Plots were rarely marked out far from railway stations, and collective farmers often used the right-of-ways as a common source of fodder. Scott Lyon, "Agricultural Observations During a Trip from Moscow to Odessa, June 1950," RG 166, Foreign Agricultural Service Narrative Reports, Box 63, Folder: Agriculture 1954–1950, NARA.

12. "U.S.S.R. Agricultural Commodity Information, October 21, 1949," RG 166, Foreign Agricultural Service Narrative Reports, U.S.S.R., Box 1000, Folder: Agriculture 1948–1949, NARA; Lyon, "Agricultural Observations."

13. Hamer, "Trip Report from Moscow to Odessa"; Burlin Hamer, "Trip Report from Moscow to Dzaudzhidau," July 1951, RG 166, Foreign Agricultural Service Narrative Reports, Box 63, Folder: Agriculture 1954–1950, NARA.

14. Zhores Medvedev, *Soviet Agriculture* (New York: Norton, 1987), 258.

15. On industrial espionage in the automobile industry, see Lewis H. Siegelbaum, *Cars for Comrades: The Life of the Soviet Automobile* (Ithaca, NY: Cornell University Press, 2011), 10–35. On technology transfer and its perceived abuse by the Soviets, see Kendall Bailes, "The American Connection: Ideology and the Transfer of American Technology to the Soviet

Union, 1917–1941," *Comparative Studies in Society and History* 23, no. 3 (July 1981): 421–48.

16. In 1955, one national kolkhoz newspaper advertised a short course to become a "certified rationalizer, . . . saving time and energy in the workplace." *Kolkhoznoe selo,* 19 April 1955, 1; Deborah Fitzgerald makes a similar argument for U.S. agriculture in the 1920s. Deborah Fitzgerald, *Every Farm a Factory: The Industrial Ideal in American Agriculture* (New Haven, CT: Yale University Press, 2003), 1–8.

17. Neil J. Melvin, *Soviet Power and the Countryside: Policy Innovation and Institutional Decay* (New York: Palgrave Macmillan, 2003), 29–44.

18. My understanding of Lysenkoism is based on Loren Graham, *Science, Philosophy, and Human Behavior in the Soviet Union* (New York: Columbia University Press, 1987); David Joravsky, *The Lysenko Affair* (Cambridge, MA: Harvard University Press, 1970); Nikolai Krementsov, "A 'Second Front' in Soviet Genetics: The International Dimension of the Lysenko Controversy, 1944–1947," *Journal of the History of Biology* 29 (1996); Dominique Lecourt, *Proletarian Science? The Case of Lysenko* (London: NLB, 1977); Zhores A. Medvedev, *The Rise and Fall of T. D. Lysenko* (New York: Columbia University Press, 1969); Nils Roll-Hansen, *The Lysenko Effect: The Politics of Science* (Amherst, NY: Humanity, 2004); and Valerii Soifer, *Lysenko and the Tragedy of Soviet Science,* trans. Rebecca Gruliow and Leo Gruliow (New Brunswick, NJ: Rutgers University Press, 1994); as well as on Lysenko's published works, including T. D. Lysenko, *The Science of Biology Today* (New York: International, 1948) and *New Developments in the Science of Biological Species* (Moscow: Foreign Languages Publication House, 1951).

19. On Fordism and Taylorism in the USSR, see Frederick Taylor, *The Principles of Scientific Management* (New York: Harper and Brothers, 1911); Bailes, "The American Connection," as well as Kendall Bailes, "Alexei Gastev and the Soviet Controversy over Taylorism, 1918–24," *Soviet Studies* (1977); and James Clay and Richard F. Vidmer Thompson, *Administrative Science and Politics in the USSR and the United States: Soviet Responses to American Management Techniques* (New York: Bergin and Garvey, 1983).

20. Loren Graham, *Science in Russia and the Soviet Union: A Short History* (New York: Cambridge University Press, 1993), 124.

21. This is similar to the distinction Mark Adams makes between science and ideology in his history of the Kol'tsov Institute. Here, I make a distinction between Lysenko and Lysenkoism. See Mark Adams, "Science, Ideology, and Structure: The Kol'tsov Institute, 1900–1970," in *The Social Context of Soviet Science,* ed. Linda Lubrano and Susan Gross Solomon (Boulder, CO: Westview, 1980), 173–204.

22. T. D. Lysenko, *New Developments in the Science of Biological Species* (Moscow: Foreign Languages Publishing House, 1951), speech at All-Union Agricultural Academy, 7 August 1948.

23. The best work on the rivalry between Vavilov and Lysenko is Roll-Hansen, *The Lysenko Effect*. My understanding of the work of Vavilov is based on the following: David Joravsky, "The Vavilov Brothers," *Slavic Review* 24, no. 3 (1965); Mark Popovsky, *The Vavilov Affair* (Hamden, CT: Archon, 1984); and I. A. Zakharov, *Nikolai Ivanovich Vavilov i stranitsy istorii sovetskoi genetiki* (Moscow: Institut obshchei genetiki im. N. I. Vavilova RAN, 2000).

24. See Popovsky, *The Vavilov Affair*; Roll-Hansen, *The Lysenko Effect*; and Zakharov, *Nikolai Ivanovich Vavilov*.

25. See Lysenko, *New Developments in the Science of Biological Species*.

26. On harvest failure, see Lecourt, *Proletarian Science*; Medvedev, *The Rise and Fall of T. D. Lysenko*; Valerii Soyfer, *Lysenko and the Tragedy of Soviet Science* (New Brunswick, NJ: Rutgers University Press, 1994); and Conrad Zirkle, *Death of a Science in Russia: The Fate of Genetics as Described in Pravda and Elsewhere* (Philadelphia: University of Pennsylvania Press, 1949). On harvest variability, see Nikolai M. Dronin and Edward C. Bellinger, *Climate Dependence and Food Problems in Russia 1900–1990* (New York: Central European University Press, 2005), 125, 194, 289.

27. Medvedev, *The Rise and Fall of T. D. Lysenko*, 238–39. See also LeCourt, *Proletarian Science*, 89; Nikolai Krementsov, "A 'Second Front' in Soviet Genetics: The International Dimension of the Lysenko Controversy, 1944–1947," *Journal of the History of Biology* 29, no. 2 (1996): esp. 249–50.

28. On cold tolerance in calves, see "The Cold Method of Rearing Calves," *Kolkhoznoe proizvodstvo*, November 1949, 27.

29. On ducks, see "Determination of Sex in Day-old Ducklings," *Kolkhoznoe proizvodstvo*, February 1949, 38–39. On ESP, see Michael Froggatt, "Science Education under Khrushchev," in *The Dilemmas of De-Stalinization: Negotiating Cultural and Social Change in the Khrushchev Era*, ed. Polly Jones (London: Routledge, 2006), 250–266. State-supported projects of dubious scientific merit were not unique to the Soviet Union. For an excellent overview of the function of pseudoscience, see Michael Gordin, *The Pseudoscience Wars: Immanuel Velikovsky and the Birth of the Modern Fringe* (Chicago: University of Chicago Press, 2012).

30. "An article from *Veterinariia*, September 1949, notes 'the basic failure in the progress of the fulfillment of the state plan of development of livestock raising in 1949 as an analysis of the information has shown, is still the low output of young stock . . . there is still a significant waste and

squandering of livestock on the farms. . . . failures in the increase of young stock are the special responsibility of the livestock specialists and especially the zoo-technicians who have paid insignificant attention to the timely and correct organization of breeding and artificial insemination [and] it is necessary to proclaim a merciless campaign with the appearance of any kind of negligent attitude towards livestock which leads to the sickness of livestock.'" Translation of article, signed Kirk, located in RG 166, Foreign Agricultural Service Narrative Reports, 1950–1951, Box 484, Folder: Livestock Industries, NARA.

31. David Hoffman, "Mothers in the Motherland: Stalinist Pronatalism and Its Pan-European Context," *Journal of Social Science* 34 (Fall 2000): 35–54.

32. V. V. Matskevich, "Extension and Farm Management," speech, May 1948, trans. Joseph Bulik, RG 166, Foreign Agricultural Service Narrative Reports, 1946–1949, Box 1005, NARA; DAKhO, f. R-6184, op. 1, d. 302.

33. My understanding of the prerevolutionary history of Askaniia Nova comes from Waldemar von Falz-Fein, *Askaniia Nova* (Kiev: Agrarna Nauka, 1997); A. D. Babich, *Stepnoi oazis: Askaniia Nova kharakteristika prirodnykh uslovii raiona* (Kharkov: Izd. Khakovskogo Universiteta im. Gorkogo, 1960); and Lisa Heiss, *Askania-Nova: Animal Paradise in Russia* (London: Bodley Head, 1970). Falz-Fein's project was smaller in scale than, but similar in approach and attitude to that practiced by acclimatization stations in European colonial territories. See Michael A. Osborne, *Nature, the Exotic, and the Science of French Colonialism* (Bloomington: Indiana University Press, 1994).

34. David Moon, *The Plough that Broke the Steppes: Agriculture and Environment on Russia's Grasslands, 1700–1914* (Oxford: Oxford University Press, 2013), 301–2; Douglas R. Weiner, "Community Ecology in Stalin's Russia: 'Socialist' and 'Bourgeois' Science," *Isis* 75, no. 4 (1984); and Douglas R. Weiner, *A Little Corner of Freedom: Russian Nature Protection from Stalin to Gorbachev* (Berkeley: University of California Press, 1999). On natural limits in the realm of Soviet forestry, see Stephen Brain, *Song of the Forest: Russian Forestry and Stalin's Environmentalism* (Pittsburgh: University of Pittsburgh Press, 2011). Soviet officials were just as likely as historians to forget the park's history of landscape management. A 1950 newspaper article from Kharkov refers to Askaniia Nova as "20,000 hectares that have never been plowed, virgin . . . steppe." *Kolkhoznoe selo,* 4 April 1950, 2.

35. For more on the poorly defined term "Lysenkoism," see Graham, *Science, Philosophy, and Human Behavior in the Soviet Union,* 124–50, and Medvedev, *The Rise and Fall of T. D. Lysenko,* 192.

36. The definition of the term "hybrid" varies. See Harriet Ritvo, *The Platypus and the Mermaid and Other Figments of the Classifying Imagination* (Cambridge, MA: Harvard University Press, 1997), 85–130. Here I use hybrid according to the definition generally accepted by plant breeders, in which two populations of the same species are crossed.

37. Loren Graham makes a similar point about the vaguely defined technique of "vernalization." Graham, *Science in Russia and the Soviet Union,* 124–25.

38. Robert Bakewell's success in sheep breeding in England is the classic example of this process. Roger J. Wood and Vitezslav Orel, *Genetic Prehistory in Selective Breeding: A Prelude to Mendel* (Oxford: Oxford University Press, 2001), 57–92.

39. RGAE, f. 8390, op. 2, d. 2794, l. 202.

40. RGAE, f. 8390, op. 2, d. 1558, l. 12.

41. "Scientist Michurin and His Work with Mixed-Race Cattle," *Kolkhoznoe proizvodstvo,* October 1948, 27–28; RGAE, f. 8390, op. 2, d. 1558, l. 12 and d. 2794, l. 202.

42. RGAE, f. 8390, op. 2, d. 1558, l. 12.

43. *Five-Year Plan for the Rehabilitation and Development of the National Economy of the USSR;* "On the Measures to Build Up Agriculture in the Post-War Period," *Pravda,* 28 February 1947; Lysenko, *New Developments in the Science of Biological Species.*

44. DAKhO, f. 4672, op. 7, d. 4004.

45. RGAE, f. 7803, op. 4, d. 1349, l. 79.

46. RGAE, f. 8390, d. 1558, op. 12; One potential indication that there were problems was a 1951 news story deprecating the work of Soviet breeders. As an American diplomat in Russia reported, "Even Soviet hog breeders are not immune to the accusation of 'dogmatism' and 'formalism' in their work—labels which definitely are not an aid to one's career in the USSR." In *Izvestiia,* 9 August 1951, Academician L. K. Greben was sharply and sarcastically criticized for his inordinate pride in considering himself the heir of the late M. F. Ivanov, noted for his development of the Ukrainian White Steppe breed of pig. This vanity allegedly led him to reject the valuable advice of his colleagues on methods of preserving and improving that breed, which had deteriorated under his direction. Greben's major crime, it seems, was to suggest putting shoes on the pigs to prevent lameness. "'It is obvious why they are lame,' concluded the HEIR, 'if you walked around barefoot as they, you too would be lame. We must make shoes for the pigs!'" Moscow Embassy to Department of State, August 30, 1951, RG 166, Foreign Agriculture Service Narrative Reports, 1950–1951, Box 484, Folder: Livestock Industries, NARA.

47. RGAE, f. 7803, op. 4, d. 1349, l. 56.

48. For example, "From Each Sow: 27 Piglets," *Kolzhoznoe proizvodstvo*, June 1949, 36–37.

49. DAKhO, f. R-6184, op. 1, d. 302, l. 3.

50. DAKhO, f. R-6184, op. 1, d. 302, l. 3.

51. DAKhO, f. R-6184, l. 4.

52. DAKhO, f. R-6184, op. 1, d. 5 an d. 7.

53. Loren Graham makes this same point, although he attaches less importance to it. Graham, *Science, Philosophy and Human Values*, 137.

Chapter 4
Substituting Meat

1. Harrison E. Salisbury, "Nixon in Wrangle with Khrushchev," *New York Times*, 25 July 1959.

2. Victoria de Grazia, *Irresistible Empire: America's Advance through Twentieth-Century Europe* (Cambridge, MA: Harvard University Press, 2005), 153–56.

3. On death and displacement of livestock, see J. A. Newth, "Soviet Agriculture: The Private Sector 1950–1959. Animal Husbandry," *Soviet Studies* 13, no. 4 (1962): 129. On canned provisions, see Edward R. Stettinius, *Lend-Lease, Weapon for Victory* (New York: Macmillan, 1944), 216–18.

4. In July 1946 in Kiev, an 800-gram can of *tushonka* cost 14 rubles in a state ration store. No fresh meat was available in these stores, but fresh pork was available in a *gastronom* for 11–15 rubles per dekagram, approximately eight times the price of the canned food. For further comparison, a plucked, undressed chicken (no weight given) was not available in the state store, and cost 130 rubles in a *gastronom*, and 85–90 rubles at a public market. L. A. Skeoch, "Food Prices and Ration Scale in the Ukraine, 1946," *Review of Economics and Statistics* 35, no. 3 (1953): 231.

5. Stettinius *Lend-Lease, Weapon for Victory*, 16.

6. "Records relating to Lend Lease with the USSR 1941–1952," RG 59, General Records of the Department of State, Box 6, Office of Soviet Union Affairs, NARA.

7. In 1945 the Leningrad meat factory canned only 1,699 tons of meat. This figure jumped in later years. RGAE, f. 8297, op. 5, d. 438.

8. M. M. Danilov, *Tovarovedenie prodovol'stvennykh tovarov: Miaso i miasnye tovarye* (Moscow: Ekonomika, 1964).

9. *Kniga o vkusnoi i zdorovoi pishche* (Moscow: AMN, 1952), 161.

10. On midcentury hog operations in the United States, see Joseph Anderson, "Lard to Lean: Making the Meat-Type Hog in Post–World War II America," in *Food Chains: From the Farmyard to the Shopping Cart*, ed. Warren Belasco and Roger Horowitz (Philadelphia: University of Pennsylvania Press, 2009), 29–46.

11. On Soviet pig types, especially the persistence of "lard and bacon" types, see the remarks of J. Marion Steddom, an Iowa hog expert, upon visiting several Soviet meatpacking facilities in 1955. "Prilozheniia k otchetu . . . ," RGAE, f. 7486, op. 22, d. 106, ll. 83–84, and "Itineraries, Notes, Memoranda of the American Delegation to the USSR," RS 21/7/65, Box 1, Folder 5, J. Marion Steddom Papers, Iowa State University Special Collections.

12. On farm consolidation, see TsDAVO, f. R-27, op. 18, d. 6845, ll. 32–45, and Roy Medvedev, *Khrushchev: The Years in Power* (New York: Norton, 1978), 86.

13. "Visits to Soviet Agricultural Installations: November 15, 1961," RG 166, Foreign Agricultural Service Narrative Reports 1955–1961, Folder: Agriculture, NARA.

14. William Taubman, *Khrushchev: The Man and his Era* (New York: Norton, 2004), 371–75.

15. "Toward an Even Higher Output of Ice Cream," *Molochnaia promyshlennost'* (July 1961): 18–21; As early as 1947 (a year of famine in the USSR) a dairy in Baku reported that one-sixth of its daily output was ice cream. "Materials, Protocols, and Reports of the Bureau of Technical Expertise for MinMiasProm (Ministry of Meat Production)," RGAE, f. 8295, op. 4, d. 198, l. 197.

16. Yu. A Olenev and N. D. Zubova, *Kak proizvodstvo morozhenoe* (Moscow: Pishchevaia Promyshlennost', 1977), 3; "Resolutions of the First All-Union Industrial Conference of the Workers of Public Nutrition," *Voprosy pitaniia* no. 6 (1933): 2. For an overview of communal feeding projects, see Susan Reid, "Cold War in the Kitchen: Gender and De-Stalinization of Consumer Taste in the Soviet Union under Khrushchev," *Slavic Review* 61, no. 2 (2002): 211–52.

17. DAKhO, f. R-4447, op. 2, d. 648, l. 23.

18. Daniel Block, "Milk," in *The Oxford Companion to American Food and Drink*, ed. Andrew F. Smith (Oxford: Oxford University Press, 2007), 385–88.

19. "New Developments in Cooling Railroad Cars," *Kholodil'naia tekhnika*, no. 3 (1960): 46–47.

20. "About Dry Ice," ibid., 43–45; for one Leningrad municipal cold locker the cost of purchasing a new freezer in 1954 was nearly half the an-

nual operating cost of the facility: 1,044.80 rubles of 2,621.20 total. RGAE, f. 9355, op. 4, d. 4, l. 8.

21. "Work on Home Refrigerators: A Report by Mechanic M. Badzhi," *Kholodil'naia tekhnika*, no. 5 (1960): 51–52.

22. Ibid.

23. N. Lyubimov and A. Burmakin, "New Equipment for Mechanization of Ice Cream Production at RosMiasorybtorg Plants," ibid., no. 3 (1960): 32. The desire for complete automation was hardly unique to the Soviet Union. See David Noble, *Forces of Production: A Social History of Industrial Automation* (Oxford: Oxford University Press, 1986), esp. 42–43.

24. "Increasing the Output of High-Quality Ice Cream," *Pishchevaia promyshlennost' (Molochnaia)*, no. 3 (1961): 11; "Development of Industrial Ice Cream Processing," *Kholodil'naia tekhnika*, no. 5 (1970): 1–3.

25. G. Azov, "New Varieties of Ice Cream," *Kholodil'naia tekhnika*, no. 1 (1960): 39–41. Recipes are given for tomato, carrot, and prune ice cream.

26. Consumer preferences were notoriously hard to identify in the Soviet Union, but at least one publication was specific about the *plombir* preference: "Increasing the Output of High-Quality Ice Cream," 11. The crème brûlée preference is my own observation based on informal discussions with residents of the former Soviet Union.

27. George Bergstrom, "The Soviet Food Front" (unpublished report, Food Science Laboratory, Michigan State University, 1959).

28. Ibid., 18.

29. Ibid.

30. "Uruguay-USSR Agriculture," RG 166, Foreign Agricultural Service Narrative Reports 1955–1961, Box 870, Folder: Agriculture, NARA. The Caspian seal has been listed as an endangered species since 2008.

31. RGAE, f. 7803, op. 1, d. 1834, l. 92–93.

32. I. Volper, *The Soviet Food Industry* (Moscow: Foreign Languages Publishing House, 1958), 11.

33. Elena Molokhovets, *Classic Russian Cooking: Elena Molokhovets' A Gift to Young Housewives*, trans. and ed. Joyce Toomres (Bloomington: Indiana University Press, 1998), 54–60; *Kniga o vkusnoi i zdorovoi pishche*, 86, 105, 192, 203, 377.

34. *Kniga o vkusnoi i zdorovoi pishche*, 105. This was signaled also by a name change, from the traditional Russian word *pomodory*, inherited from its Turkish importers (who in turn got it from the Italian *pomodoro*), to *tomaty*, the more commonly occurring name across Europe.

35. Reid, "Cold War in the Kitchen," 211–12.

36. Piglet rustling was discussed by future Soviet Minister of Agri-

culture Vladimir Matskevich in his 1947 speech to the Central Committee. There were legal systems of animal distribution as well. Between 1945 and 1949, baby animals were given to star workers as a reward for good work in place of monetary wages.

37. For illustrations of postwar kitchen designs, see *Kniga o vkusnoi i zdorovoi pishche*, 33, as well as D. Tsolov and R. Krustanova, *Mebeli: Eksperimentalni obraztsi za masovo proizvodstvo na spal'ni i kukhn*, (Sofia: Izd-vo na Bŭlgarskata akademiia na naukite, 1957).

38. Ibid.

Chapter 5
The Old and the New

1. P. M. Zemskii, *Razvitie i razmeshchenie zemledeliia po prirodno-khoziaistvennym raionam SSSR* (Moscow, Gosudarstvennoe Izdatel'stvo, 1959), 154.

2. For an overview, see Stephan G. Prociuk, "The Territorial Pattern of Industrialization in the USSR: A Case Study in Location of Industry," *Soviet Studies* 13, no. 1 (1961): 69–95.

3. Steven Marks, *Road to Power: The Trans-Siberian Railroad and the Colonization of Asian Russia, 1850–1917* (Ithaca, NY: Cornell University Press, 1991). See also Brian Boeck, "Containment vs. Colonization: Muscovite Approaches to Settling the Steppe," in *Peopling the Russian Periphery: Borderland Colonization in Eurasian History*, ed. Nicholas Breyfogle, Abby Schrader and Willard Sunderland (London: Routledge, 2007), 41–60; A. A. Dolgoliuk, *Formirovanie trudovykh kollektivov Bratsko-Ust'-ilimskogo TPK 1955–1980* (Novosibirsk: Nauka, 1988), 62; Zemskii, *Razvitie i razmeshchenie*,147.

4. Lyudmila M. Saburova, *Kul'tura i byt russkogo naseleniia priangar'ia* (Leningrad: Nauka, 1967); RGAE, f. 7468, op. 15, d. 263, l. 120; GAIO, f. R-147, op. 1, d. 1628 (a). Tractor numbers are more difficult to gauge, since tractors came in many different sizes at this time. The small kolkhoz of Bolshevik had forty adults and one tractor (a tiny DT-14) in 1958. In 1960 the kolkhoz Taezhnik had eighty-eight adults and two tractors, both medium-sized DT-54s. GAIO, f. R-147, op. 1, dd. 1309 and 1313.

5. Marks, *Road to Power*, 53–59.

6. Gosudarstvennyi arkhiv noveishei istorii Irkutskoi oblasti (GA-NIIO), Irkutsk, f. 127 op. 27, d. 10.

7. Leonid Bezrukov, *Angara: Doch' Baikala*, (Irkutsk: Uliss, 1994).

8. Elena Shulman, *Stalinism on the Frontier of Empire: Women and*

State Formation in the Soviet Far East (Cambridge: Cambridge University Press, 2008), 11–24; Michaela Pohl, "The Virgin Lands Between Memory and Forgetting: People and Transformation in the Soviet Union, 1954–1960" (Ph.D. diss., Indiana University, 1999), chap. 2.

9. On hardships experienced during the Virgin Lands Campaign, see Pohl, "The 'Planet of 100 Languages': Ethnic Relations and Soviet Identity in the Virgin Lands," in Breyfogle et al., *Peopling the Russian Periphery,* 238–62. On conditions in Bratsk, see "Visit to Irkutsk, Bratsk and the Lake Baikal Area," 21 December 1961, RG 166, Foreign Agricultural Service Narrative Reports 1955–1961, Box 870, NARA; on statistics for baths, dugouts, and tents, see GAIO, f. R-127, op. 27, d. 12, l. 5.

10. "Visit to Irkutsk, Bratsk and the Lake Baikal Area."

11. Valentin Rasputin, *Farewell to Matyora* (Chicago: Northwestern University Press, 1995); Saburova, *Kul'tura i byt.*

12. Michel Aloys and Stephen Klain, "Current Problems of the Soviet Electric Power Industry," *Economic Geography* 40, no. 3 (1964): 206–20; Prociuk, "The Territorial Pattern of Industrialization in the USSR."

13. Prociuk, "The Territorial Pattern of Industrialization in the USSR," 74.

14. RGAE, f. 616, op. 1, d. 5672; GAIO, f. R-147, op. 1, d. 1309, l. 16.

15. RGAE, f. 7803, op. 1, d. 1861, ll. 41–42.

16. Gosudarstvennyi arkhiv rossiiskoi federatsii (GARF), Moscow, f. 616, op. 1, d. 5642.

17. N. N. Kazanski, "Selsko-khoziaistvennoe osvovenie srednogo priangaria v sviazi so stroitelstvym Bratskoi GES," *Izvestiia* 7 (1960): 53–60.

18. RGAE, f. 7803, op. 1, d. 1861, l. 38; "Visit to Irkutsk, Bratsk and the Lake Baikal Area."

19. Dmytryshyn, *The History of Siberia,* 22–24.

20. Janet Martin, *Treasure of the Land of Darkness: The Fur Trade and Its Significance for Medieval Russia* (Cambridge: Cambridge University Press, 1986). 144–46; John F. Richards, *The Unending Frontier: An Environmental History of the Early Modern World* (Berkeley: University of California Press, 2003), 531–40.

21. Richards, *The Unending Frontier,* 531–40.

22. Saburova, *Kul'tura i byt.*

23. Dmytryshyn, *The History of Siberia,* 22, 31–32.

24. Saburova, *Kul'tura i byt.*

25. N. N. Bakeyev and A. A. Sinitsyn, "Status and Conservation of Sables in the Commonwealth of Independent States," in *Martens, Sables, and Fishers: Biology and Conservation,* ed. S. W. Buskirk, A. S. Harestad, M. G. Raphael,

and R. A. Powell (Ithaca, NY: Cornell University Press), 246–54; John Long, *Introduced Mammals of the World; Their History, Distribution and Influence* (Cambridge, MA: CABI, 2003), 291–92; GAIO, f. R-147, op. 1, d. 10, l. 21.

26. Piers Vitebsky, *Reindeer People: Living with Animals and Spirits in Siberia* (London: Harper, 2005), 40–62.

27. Yuri Slezkine, *Arctic Mirrors: Russia and the Small Peoples of the North* (Ithaca, NY: Cornell University Press, 1994), 150–86.

28. Aleksandr Pika and Dmitri Bogoiavlenskii name this *lumpenization*. Aleksandr Pika, ed., *Neotraditionalism in the Russian North: Indigenous Peoples and the Legacy of Perestroika* (Seattle: University of Washington Press, 1999), 8, 28–29; RGAE, f. 7803, op. 1, d. 1862, ll. 57–58.

29. "Fur: Soft Gold," *Vostochno-Sibirskaia pravda*, 14 January 1953, 4.

30. Pika, *Neotraditionalism in the Russian North,* 98–100.

31. The fur industry opposed such legislation. The *New York Times* reported that "surveys made by fur publications indicate that not more than one out of ten women have voiced any interest in the country of origin in connection with the furs that they wear." "U.S. Fur Dealers Will Visit Leningrad for Auctions if the Soviet Grants Visas," *New York Times*, 19 May 1951; Pika, *Neotraditionalism in the Russian North,* 98–99.

32. RGAE, f. 7803, op. 1, d. 1861, ll. 114–19.

33. RGAE, f. 7803. op. 1. d. 1834 l. 19, and d. 1862, l. 57.

34. GAIO, f. R-147, op. 1, d. 1313 .

35. GAIO, f. R-147, op. 1, d. 1313, ll. 14–15.

36. V. H. Beregovoy and J. Moore Porter, *Primitive Breeds—Perfect Dogs* (Arvada, CO: Hoflin, 2001), 424; V. G. Gusev, "*Okhota s laikoi,*" *Fizkultura i sport* (Moscow, 1978).

37. A. T. Voilochnikov and S. D. Voilochnikov, *Okhotnichyi laiki* (Moscow: Lesnaia Promyshlennost', 1982).

38. Donald Treadgold, *The Great Siberian Migration: Government and Peasant in Resettlement from Emanicpation to the First World War* (Princeton, NJ: Princeton University Press, 1957).

Epilogue

1. Interview, Samuel Fry, 26 January 1993, Association for Diplomatic Studies and Training, Foreign Affairs Oral History Project, http://adst.org/oral -history/oral-history-interviews/#f.

2. John F. Kennedy Presidential Library, Boston, Folder: USSR: Wheat Sale, 6.

3. E. Strauss, "The Soviet Dairy Economy," *Soviet Studies* 21, no. 3 (January 1970): 269–73.

4. Alec Nove, "Soviet Agriculture under Brezhnev," *Slavic Review* 29, no. 3 (September 1970): 379–410.

5. Alexander Gerschenkron, *Economic Backwardness in Historical Perspective* (Cambridge, MA: Harvard University Press, 1962).

Glossary of Russian Terms

Doiarka:	milkmaid
Gibrid:	hybrid
Gosplan:	State Planning Committee. Responsible for the Five Year Plans
Iasak:	tribute, usually pelts, demanded from indigenous Siberians by the Russian imperial authorities
Kolkhoz:	collective farm, the smaller state-owned unit of agriculture
Krug:	Large administrative territory, in Soviet times the word was interchangeable with "oblast."
Kulak:	wealthier peasants, scapegoated as capitalist wreckers during the early 1930s
Laika:	Siberian husky dog (literally in Russian: "barker")
Metis:	mixed-breed (from the French *métis*)
Minselkhoz (*Ministerstvo Sel'skogo Khoziaistva*):	Soviet Ministry of Agriculture
NEP (*Novaia Ekonomicheskaia Politika*):	New Economic Policy that encouraged private ownership and capital investment in the early Soviet Union

Oblast:	Large administrative territory, in some ways equivalent to an American state
Podzol:	poor agricultural soil typical of Central and Eastern Siberia
Porodnost':	purebred status
Promyshlennik:	a provisioner or middleman in the Siberian fur trade between indigenous groups and the Russian imperial government
Raion:	Smaller administrative territory, in some ways equivalent to an American county
Salo:	lightly cured pork fat

Index

Illustrations, maps, and notes are indicated by "f," "m," and "n" following the page numbers.

Acclimatization projects, 108, 127, 129, 130–131, 133

Agricultural initiatives: animals, 106–150; food production, 151–187; kolkhozes (collective farms), 21–62; modernization. See Agricultural modernization; Quarantine Stations, 83–105; sovkhozes (state-owned farms), 66–67, 71–72, 176–177, 198–199, 201–202, 206

Agricultural modernization: animal agriculture, 107, 111–112, 116–119, 127, 149–150; deferral to outside authorities, 57; environmental effects of, 12–13, 107, 228–229; failures of, 1, 19–20; gaps between plans and reality in, 1, 2, 15, 17–18, 21, 227–228; lessons learned, 229–232; mechanization, 7–8, 29, 59–60; natural environment as barrier to, 35; public visibility of, 229, 230–231; in rural areas, 9–10, 21–22, 119; ulterior motives of, 228; and "will to improve," 34. See also Collectivization

Agrobiology, 123. See also Lysenkoism

Agro-ecological perspective, 11

Alfalfa, 58, 59

Allotment gardens, 67–68, 69, 102

All-Union Agricultural Academy, 123

Amerikanka enclave for engineers, 238n17

Angara River, 190, 194–195, 200

Animal agriculture, 106–150; adaptations to local environment, 114; and collective ownership, 113; disease concerns, 108–109; for draft labor, 110, 111; and food

Animal agriculture (*continued*)
    scarcity, 109–110; industrial-
    ization and modernization
    of, 107, 111–112, 116–119, 127,
    149–150; lack of productivity
    in, 226; on model farms vs.
    countryside, 114–117; post-
    war recovery of, 108–109,
    148–149; and private owner-
    ship, 112–113; pronatalist pol-
    icies of, 127–128, 135; show-
    manship in, 108; socialist
    approach to, 108; U.S. per-
    ceptions of, 114–116. *See also*
    Breeding; Lysenkoism; *spe-*
    cific animals
Antibiotics, 13; lack of, 17, 107
Anti-Semitic campaigns, 69–70
Apartment life, 182, 183–184
Architectural designs influencing
    food production, 181, 183–184
Askaniia Nova breeding station,
    129–133, 135–137, 143–144,
    247*n*33
Austria, potato beetles in, 90

Barnyards, reorganization of, 161–162
Beef cattle, 116
Beetles. *See* Potato beetles
Belaia Dacha farm, 176–177
Big type pigs, 160
Bolshevik farm, 201, 218, 252*n*4
Bolsheviks, 2–3, 7–8, 25
Book of Delicious, Healthy Food
    (Ministry of Health and
    Food Provisioning), 177, 179
Bratsk dam project, 190, 193, 194,
    197–201
Breeding: challenges in, 113; cows,
    progression from mixed-

breed to purebred status,
    142–146; female swineherds,
    role of, 137–141, 142*f*; of laika
    dogs, 221; Lysenkoist meth-
    odology for, 17, 126–129,
    132–133, 135–136, 138–143;
    pigs, debates on, 160–161;
    state farms for, 127–129,
    131–133, 135, 136–137; of
    Ukrainian White Steppe
    pig, 133–136, 139, 141,
    248*n*46. *See also* Askaniia
    Nova breeding station
Bukharin, Nikolai, 26
Bulik, Joseph, 75, 76
Burds, Jeffrey, 92
Buriats, 194–195, 207
Bush, Guy: agricultural background
    of, 24; on challenges in kol-
    khozes, 14–15, 35, 61; disil-
    lusionment experienced by,
    56–58; duties and responsi-
    bilities of, 24, 30; evidence
    supporting observations of,
    33–34; on feed for animals,
    50–51, 81; on health concerns
    and disease in animals, 53;
    on housing for animals, 51;
    interpreters utilized by, 55; on
    leadership at kolkhozes, 42;
    on living and working con-
    ditions, 34, 38, 39, 43, 44; on
    management and workforce
    ineptitudes, 46, 47; preju-
    dices toward Soviet Union,
    33; recommendations for
    change in pig farming prac-
    tices, 40–41; return to U.S.,
    54. *See also* Millerovo farm
Butter, distribution of, 167

Cafeterias, 152, 182–183
Canned foods industry, 154, 155–156,
    178–180. *See also Tushonka*
Capitalism: fences as tools of, 52;
    and industrial efficiency, 7,
    25–26; persecution of farm-
    ers practicing, 27–28; tech-
    niques adopted by Soviet
    Union, 120, 121
Caspian seal, 176, 251n30
Cattle. *See* Cows and cattle
Central Committee: collectiviza-
    tion, implementation by,
    26–27; on foreign experts at
    kolkhozes, 54–55; funding
    for Quarantine Stations, 90
Chloropicrin, 87, 97–98
Cholera, 53, 83–84
Chukchi, 222
Citrus industry, 154
Climate, as environmental chal-
    lenge, 11, 35, 59
Cold War, impact on development
    and rebuilding efforts, 9–10,
    126, 154
Collective farm revolts (1930), 68
Collectivization: challenges of,
    14–15; consequences of, 60;
    enforcement through vio-
    lence and persecution,
    27–28; failures of, 21–22,
    60–61; grain production
    and distribution impacted
    by, 28–29; historical back-
    ground of, 25–26, 237n6;
    implementation of, 4, 26–27;
    inducements for, 28; as
    method of control, 4, 60;
    resistance to, 27–28, 61, 68.
    See also Kolkhozes

Colorado potato beetles. *See*
    Potato beetles
Committee of the North, 213–214
Communism: rural resistance to,
    22; service of youth as devo-
    tion to, 199; as urban ideol-
    ogy, 3
Consumption patterns, changes in,
    153–154, 178, 181, 186
Control: collectivization as
    method of, 4, 60; human
    control over nature, 8–9,
    23–24, 35; Quarantine Sta-
    tions as method of, 87–88,
    90–91, 98, 102–103
Cookbooks, 177–178, 179, 187
Corn, as feed crop, 162–163, 164, 226
Countryside. *See* Rural governance
Cows and cattle: adaptations to
    local environment, 114; beef
    cattle, 116; dairy cows, 116,
    141–143, 146–147, 226; free
    grazing of, 115; progres-
    sion from mixed-breed to
    purebred status, 142–146;
    seasonal feeding regimes,
    109–110. *See also* Animal
    agriculture; Meat produc-
    tion; Milk production
Cultural norms influencing food
    production, 181–183
*Culture and Everyday Life in a Rus-
    sian Settlement on the Ang-
    ara* (Saburova), 200

Dairy cows, 116, 141–143, 146–147,
    226
Dairy trusts, 112–113
Dam project. *See* Bratsk dam
    project

Dekulakization, 27–28, 65
Desertification, 12
Dietary concerns. *See* Public nutri-
    tion campaigns
Disease: cholera, 53, 83–84; in pigs,
    50, 52–53; postwar com-
    municable animal diseases,
    108–109; in potatoes, 15–16,
    86–87, 88–90, 91, 97–100
Distribution: collectivization
    impacting, 28–29; of dairy
    products, 167; processed
    foods as solution to prob-
    lems of, 151–152, 155, 166;
    urban bias in networks of,
    180–181, 186
Distribution rings, 202–205, 206
Doiarki (milkmaids), 142–143,
    146–147
Draft animals, 3, 110, 111
Drought, 74, 75, 77
Dry ice manufacturing, 169–170

Ecological specificity, 135–136
Energy sources, harnessing and
    refining of, 154–155
Engerman, David, 47
Environmental challenges: agricul-
    tural modernization, effects
    of, 12–13, 107, 228–229; cli-
    mate, 11, 35, 59; desertifica-
    tion, 12; diversity of regions,
    11–12; drought, 74, 75, 77;
    erosion, 12; at kolkhozes, 50,
    61; to pig farming, 50; pol-
    lution, 9, 12; water and soil
    contamination, 12. *See also*
    Disease
Ermine pelts, 215, 216
Erosion, 12

Eskimo-Generators, 172
Evenks, 207, 222

Factory farming system, 156
Fainsod, Merle, 34, 42, 44
Falz-Fein, Friedrich, Jr., 130–131,
    132, 134, 247n33
Falz-Fein, Friedrich, Sr., 129–130
Famine: of 1921 and 1922, 3, 68;
    of 1932–34, 21, 60, 77; of
    1946–47, 74–76, 77; govern-
    ment manipulation of, 22,
    60, 70, 74–77; Irish potato
    famine (1840s), 86; and state
    grain requisitions, 74, 77;
    U.S. neglect in recognizing,
    75–76; wheat exportation
    during, 10
Farewell to Matyora (Rasputin),
    200
Feed crops, 162–163, 164
Feed for animals, 50–51, 81–82,
    109–110, 162–164
Females. *See* Women
Fertilizers, 12–13
Firearm ownership policies, 217–218
Fish-canning industry, 174–175
Five Year Plans, 27, 29–30, 56, 111,
    137–138, 200–201
Flax crops, 36
Fodder crops, 162–163, 164
Food distribution rings, 202–205,
    206
Food production, 151–187; archi-
    tectural designs influencing,
    181, 183–184; consumption
    patterns, changes in, 153–154,
    178, 181, 186; creativity and
    flexibility in, 164, 169, 174;
    cultural norms influencing,

181–183; fruit and vegetable canning, 178–180; handmade products, aversion to, 159, 171–172; hygiene considerations, 159, 170, 171–172; and packaging, 185; processed foods as solution to distribution problems, 151–152, 155, 166; and public nutrition campaigns, 10–11, 154, 177, 187; and purchasing process at shops, 184–185; reform of, 152–153, 168, 181, 185–186; substitution products, 18; technological innovations in, 168–172, 180; urban bias in distribution networks, 180–181, 186; wild foods, 175–177. *See also* Famine; Meat production; Milk production; *specific foods*

Ford, Henry, 7–8, 17

Fordism, 17, 120, 121

Foreign Agricultural Service, U.S., 74–75

Foreign experts at kolkhozes, 22, 23, 29, 33, 54–55. *See also* Bush, Guy; Heikens, George

Fox pelts, 215, 216

France, potato beetles in, 90

Freezer technology, 170–171

Frozen food industry, 154. *See also* Ice cream

Fruits and vegetables, canning of, 156, 168, 178–180

Fur trade, 207–211, 212–213, 215–216, 219–222

Gagarin, Yuri, 36

Gardens, private, 67–68, 69, 102

Germany: agricultural quarantine facilities in, 84; potato beetle-breeding facilities established by, 89; Ukraine invasion and occupation by, 66, 67, 88

Germination rates, 100

Gosplemsovkhozes. *See* State breeding farms

Grain: production and distribution impacted by collectivization, 28–29; Soviet purchases from world market, 6, 19, 225–226; state requisitions of, 74, 77. *See also specific types of grains*

Gray squirrel pelts, 209–210

Grazing: in Askaniia Nova environment, 131; and corn harvests, 163; expense for Irkutsk agricultural economy, 204; and railroad corridors, 115, 244n11

Great Britain: industrialization of animal agriculture in, 112; quarantine facilities in, 84

Greben, L. K., 248n46

Growth hormones, 13

Gun ownership policies, 217–218

Gzhatsk District of Western Oblast, 36–37

Handmade food products, aversion to, 159, 171–172

Harriman, Averill, 75–76

Heikens, George: agricultural background of, 24; on animal feed, 81; on animal health and disease, 53; on animal housing, 53; criti-

Heikens, George (*continued*)
    cism aimed at, 47–48; disil-
    lusionment experienced by,
    56–58; duties and respon-
    sibilities of, 24, 30, 37; evi-
    dence supporting obser-
    vations of, 33; interpreters
    utilized by, 55; on kolkhoze
    challenges, 14–15, 35, 39, 61;
    on kolkhoze leadership, 42;
    on living and working con-
    ditions, 34, 38–39, 43–44;
    on management and work-
    force ineptitudes, 44–47; pig
    farming recommendations
    of, 40, 41; prejudices toward
    Soviet Union, 33; return
    to U.S., 54. *See also* Rodo-
    manovo farm
Historical background: of collec-
    tivization, 25–26, 237n6; of
    Irkutsk Oblast, 194–196; of
    Quarantine Stations, 83–84;
    of Soviet regime, 2–6
Home kitchens, 183–184
Home refrigerator-freezers, 170–171
Hot-blooded pigs, 160
Housing considerations in pig
    farming, 51–53, 161–162
Human control over nature, 8–9,
    23–24, 35
Hunting, 207–223; as economic
    livelihood, 18–19, 207–208;
    and fur trade, 207–211,
    212–213, 215–216, 219–222;
    and gun ownership poli-
    cies, 217–218; laika dogs
    used for, 220–221; Ministry
    of Agriculture sponsored
    programs, 189, 207, 218–219;

    and Sibiriak identity, 211–212,
    220, 221–223
Hybrid, misuses of term, 132–133,
    142
Hybridization projects, 108, 134–135
Hydroelectric dams. *See* Bratsk
    dam project
Hygiene considerations in food
    production, 159, 170, 171–172

Iasak (tribute payments), 207,
    210–211
Ice cream, 152, 164–166, 168–174,
    180
Industrial efficiency, 7–8
Interspecific hybridity, 133, 134
Intraspecific hybridity, 133
Iowa farming and farmers, 24, 40,
    54, 162, 163
Irish potato famine (1840s), 86
Irkutsk Oblast, 188–224; agricul-
    tural reforms in, 192–193,
    206; Bratsk dam project,
    190, 193, 194, 197–201; fac-
    tors limiting agricultural
    production in, 189, 195–197,
    203–204; food provision-
    ing in, 202–205; fur trade
    in, 207–211, 212–213, 215–
    216, 219–222; history of agri-
    culture in, 194–196; immi-
    gration and settlements in,
    193–195, 197–198, 203; indus-
    trial development in, 190,
    192, 193, 206, 209; kolkhozes
    in, 188–189, 196, 201–202,
    205–206, 213–214; map of,
    191*m*; natural resources in,
    194, 206–207; podzol soil in,
    196–197; Sibiriak identity in,

211–212, 220, 221–223; sov-khozes in, 198–199, 201–202, 206; technological development in, 196; Territorial Production Complexes, 190, 192, 193, 197, 223; Virgin Lands Campaign, 5, 189, 192–193, 197, 223. *See also* Hunting

Kaplan, Cynthia, 103
Kazakhstan: industrial development in, 190, 198; podzol soil in, 196–197; Virgin Lands Campaign, 5, 192–193
Kharkov Oblast: dairy product distribution in, 167; private gardens in, 102; Quarantine Station inspection sites in, 92–93
Khrushchev, Nikita: agricultural reforms implemented by, 5–6, 106–107, 189; on corn as feed crop, 162–163; on food production, 152, 153, 156; grain purchase from U.S., 226; single-family apartments, creation of, 182; on small-group work, 219
Kiev Institute of Hygiene, 97
"Kitchen Debate" (1959), 153
Kitchens, 152, 183–184
Koch, Robert, 83
Kolkhozes (collective farms), 21–62; bureaucratic management of, 22–23; challenges in, 15, 23, 35, 39, 49–50; consolidation of, 161; decline of, 66–67; efforts to control nature on, 23–24, 35; environmental challenges at, 50, 61; failures of, 35–36, 57–58, 63–64; foreign experts at, 22, 23, 29, 33, 54–55; gaps between plans and reality in, 15, 54–55, 61–62; inducements given to, 28; in Irkutsk Oblast, 188–189, 196, 201–202, 205–206, 213–214; leadership, ineffectiveness of, 42; living and working conditions, 34, 38–39; mechanization of agriculture on, 29, 59–60; postwar rebuilding of, 65, 70–71; project management methods, 48–49; reforms in labor and land use, 119; sovkhozes vs., 71–72; specialization of, 29, 30; U.S. model of farming, attempts to export, 58–59; weed infestations, 72–73. *See also* Collectivization; Millerovo farm; Rodomanovo farm; Rural governance
Kolkhoz 22. *See* Millerovo farm
Kotov, Aleksandr, 38, 42–43, 46–47
Kulaks (wealthy peasants), 27–28

Laika dogs, 220–221
Lebedinskii cattle, 144–145
Lend-Lease program, 78, 157–158
Lenin, Vladimir, 3
Lenin Agricultural Academy, 126–127
Leptinotarsa decemlineata. See Potato beetles
Lewin, Moshe, 29–30
Literacy rates, 3
Livestock. *See* Animal agriculture

Local production goals, 202

Lysenko, Trofim: appointment as director of All-Union Agricultural Academy, 123; crop research, 123; downfall, 125–126; political ambitions, 121–122; on relationship between science and farming, 123; state support for, 8–9, 124. *See also* Lysenkoism

Lysenko Breeding Station. *See* Askaniia Nova breeding station

Lysenkoism: and acclimatization projects, 108, 127, 129, 133; on breeding and handling practices, 17, 126–129, 132–133, 135–136, 138–143; criticisms of, 125–126; on ecological specificity, 135–136; on genetics and environment, 122–123, 126; and hybridization projects, 108, 134–135; longevity of, 148, 230; as management ideology, 16–17, 107–108, 119–120, 124–125; repercussions of, 229–230; scientific terms, misuse of, 132–133, 142; as state policy, 121–122

Machine Tractor Stations (MTS), 70, 103–105, 240–241n9

Maps: Irkutsk Oblast, 191m; Millerovo farm, 32m; Rodomanovo farm, 31m; Rostov Oblast, 32m; Western Oblast, 31m

Marshall Plan, 77

Matskevich, V., 128

Meat production: fish-canning industry, 174–175; gaps between plans and reality in, 17–18; and public nutrition campaigns, 10–11, 154; reform of pig farming practices to increase, 159–164; sausages, 158, 159; as symbol of national strength, 155; *tushonka*, 151, 152, 156–158, 163–164, 180, 249n4. *See also* *specific animals*

Mechanization of agriculture, 7–8, 29, 59–60

Medvedev, Zhores, 78–79, 80–81, 125–126

Mendel, Gregor, 122–123, 125, 133

Metis (mixed-breed animals), 142, 143, 145

Mikoian, Anastas, 152–153

Milk cows. *See* Dairy cows

Milkmaids, 142–143, 146–147

Milk production: and distribution process, 167; gaps between plans and reality in, 17–18; ice cream, 152, 164–166, 168–174, 180; milkmaids' role, 142–143, 146–147; and public nutrition campaigns, 10–11, 154; refrigeration capacity, lack of, 166; as symbol of national strength, 155. *See also* Cows and cattle

Millerovo farm: agriculture prior to collectivization, 30, 37; environmental challenges at, 50, 61; feed for animals, 50–51; health concerns and disease at, 53; housing for

animals, 51–52; leadership,
ineffectiveness of, 42; living
and working conditions, 38,
39, 43, 44; management and
workforce, 46, 47; map of,
32*m*; recommendations for
change in pig farming practices, 40–41
Ministry of Agriculture: animal
agriculture, moderniza-
tion of, 116–119, 127; famine cover-up by, 74; hunting programs sponsored by,
189, 207, 218–219; mecha-
nisms for restoring authority in postwar era, 64–66;
on national ecology, 88; on
pig breeding, 161; postwar
rebuilding of kolkhozes,
65, 70–71; regionally specific plants and animals,
development of, 12; on tractor restoration, 111; on wild
foods, 176, 177. *See also*
Quarantine Stations
Ministry of Food Provisioning,
167–168, 180
Ministry of Procurements, 74
Ministry of the Interior, 217–218
Mironenko, Lidiia, 94, 96–97
Mixed-breed, progression to purebred status, 142–146
Model farms. *See* Kolkhozes
Modernization. *See* Agricultural
modernization
MTS. *See* Machine Tractor Stations

Narkomzem. *See* Ministry of Agriculture
Natural limits theory, 131

Nature, human control over, 8–9,
23–24, 35
Nenets, 222
New Economic Policy (NEP),
25–26, 36–37
Nuclear arms race, 9
Nutria pelts, 215, 216
Nutrition, public campaigns to
improve, 10–11, 154, 177, 187

Packaging of food, 185
People's Commissariat of Agriculture. *See* Ministry of Agriculture
Peppers, canning of, 180
Persecution of kulaks, 27–28
Pesticides, 12–13
Petroleum-based fertilizers, 12–13
Pigs and pig farming: adaptations
to local environment, 114;
advantages of, 50, 159–160;
breeding debates, 160–161;
disease concerns, 50, 52–53;
environmental challenges
to, 50; feed considerations,
50–51, 81–82, 109–110,
162–164; female swine-
herds' role, 137–141, 142*f*;
free-ranging lifestyle of,
115; genetic stock, reorganization of, 160–161; housing considerations, 51–53,
161–162; mechanization of,
162; phenotypically distinct
groups of, 160; recommen-
dations for change, 40–41;
reforms to increase meat
production, 159–164; roast
suckling pig, 177; Ukrainian
White Steppe pig, 133–136,

Pigs and pig farming (*continued*)
    139, 141, 248*n*46. *See also*
    Animal agriculture; Meat
    production; Millerovo farm;
    Rodomanovo farm
Plant Research Station, 86
Plombir ice cream, 173
Podzol soil, 196–197
Poland, potato beetles in, 90
Pollution, 9, 12
Porodnost' (purebred status), 143,
    145
Potato beetles, 88–90, 91, 98–100
Potato blight, 86–87, 88, 90, 97–98,
    99–100
Potatoes: advantages of grow-
    ing, 15, 79–80, 81–82; dis-
    eases and predators affect-
    ing, 15–16, 86–87, 88–90,
    91, 97–100; growing con-
    ditions for, 78; lack of state
    collection system for, 81;
    manual cultivation of, 79;
    nutritional value of, 80;
    Quarantine Stations moni-
    toring of, 15–16, 86–87, 88,
    93, 97–103; storage of, 80; as
    subsistence food, 77–79
Poverty, 36, 110, 214
Private production: allotment gar-
    dens, 67–68, 69, 102; and
    animal agriculture, 112–113;
    policies and restrictions
    on, 3–4, 68–69; as wartime
    emergency provision, 69
Processed foods, as solution to dis-
    tribution problems, 151–152,
    155, 166
Promyshlenniki (professional trad-
    ers), 195, 208, 211

Pronatalism, 127–128, 135
Public nutrition campaigns, 10–11,
    154, 177, 187
Purchasing process at shops,
    184–185
Purebred status, progression from
    mixed-breed status, 142–146
Pushcart ladies, 170

Quarantine Stations, 83–105;
    bureaucratic structure of,
    84–85; comparison with
    Machine Tractor Stations,
    103–105; creation of, 15–16,
    83; demonstration projects
    showing potential of, 87;
    employees and work culture
    at, 93–95, 96–98, 243*n*39;
    historical background
    of, 83–84; ineffectiveness
    against threats to productiv-
    ity, 90, 100; inspection sites
    for, 92–93; locations of, 91;
    as method of control, 87–88,
    90–91, 98, 102–103; potatoes
    monitored by, 15–16, 86–87,
    88, 93, 97–103; public edu-
    cation events hosted by, 99;
    role of, 70, 84–85

Railways: right-of-ways, grazing
    animals near, 115, 244*n*11;
    Trans-Siberian, 190, 196
Rasputin, Valentin, 200
Raviolis, 158, 159
Red Army, 4, 64, 88
Refrigeration technology, 168–169,
    171
Reindeer, 213–214
Rickets, 53

Rodomanovo farm: agriculture prior to collectivization, 30, 36; environmental challenges at, 50, 61; failures at, 47–48; health concerns and disease, 53; housing for animals, 52, 53; leadership, ineffectiveness of, 42; living and working conditions on, 38–39, 43–44; management and workforce, 44–47; map of, 31*m*; recommendations for change in pig farming practices, 40, 41

Rostov Oblast, 30, 37, 55; map of, 32*m*

Rural governance, 63–77; famine as method of control, 22, 60, 70, 74–77; Machine Tractor Stations, 70, 240–241*n*9; neglect during Second World War, 67; policing and surveillance techniques, 73; postwar approach to, 64–66; private production, restrictions on, 68–69. *See also* Collectivization; Quarantine Stations

Russell, Edmund, 92

Russia. *See* Soviet Union

Sable pelts, 207, 209–211, 212–213, 215–216, 219–222

Saburova, Liudmila M., 200

Salo (cured pork fat), 114

Samoyed dogs, 221

Sausages, 158, 159

Scours disease, 53

Second World War, impact on Soviet Union, 4, 64–65, 67–70

Sheep, 109–110, 114

Shelter belts, 8

Showmanship in animal agriculture, 108

Siberia: ecosystems of, 208–209; hunting as economic livelihood in, 18–19, 207–208; immigration to, 193–194; industrial development in, 190, 198; overinvestment in, 223–224; Virgin Lands Campaign, 5, 189, 192–193, 197, 223. *See also* Irkutsk Oblast

Siberian huskies, 221

Sibiriak identity, 211–212, 220, 221–223

Simmental cattle, 143–144, 145

Single-family apartments, 182, 183–184

Small-group work, 219

Smetana (sour cream zone), 167

Socialism: animal agriculture, approach to, 108; on communal consumption of meals, 165, 182–183; establishment of, 25; and food collection networks, 166, 186; and industrial efficiency, 7; luxuries and conveniences as proof of success of, 153

Soil contamination, 12

Solanum tuberosum. See Potatoes

Sour cream, distribution of, 167

Soviet farms: capitalist practices of, 25–26; environmental challenges of, 11–14; historical background of, 2–6; progress and modernization

Soviet farms (*continued*)
of, 6–11; technological lags
as advantages to, 12–13,
231–232. *See also* Agricultural modernization; Ani-
mal agriculture; Collectivization; Kolkhozes; Rural
governance; Sovkhozes; *specific farms*
Soviet Institute for the Hunting
Industry, 221
Soviet Union: Cold War, impact on
development and rebuild-
ing efforts in, 9–10, 126,
154; demographic and top-
ographical features, 3; eating patterns in, 182–183, 186;
grain purchases from world
market, 6, 19, 225–226; potatoes as staple food in, 15;
propaganda against ideo-
logical enemies and agri-
cultural pests, similari-
ties between, 91–92; Second
World War, impact on, 4,
64–65, 67–70; U.S. revocation of food aid to, 78
Sovkhozes (state-owned farms):
Belaia Dacha farm, 176–177;
growth of, 66–67; in Irkutsk
Oblast, 198–199, 201–202,
206; kolkhozes vs., 71–72
Space race, 9
Spam (canned meat), 151, 157
Specialized agriculture, 29, 30
Squirrel pelts, 207, 209–211, 212,
215–216, 219–220
Stalin, Joseph: anti-Semitic campaigns instigated by, 69–70;
on collectivization, 26,

237*n*6; death of, 5, 106;
efforts to control nature, 8,
23; famine, manipulation of,
22, 60; and food production
reform, 152–153
Starvation. *See* Famine
State breeding farms, 127–129,
131–133, 135, 136–137. *See
also* Askaniia Nova breed-
ing station
State grain requisitions, 74, 77
State-owned farms. *See* Sovkhozes
State terrorism, as method to
enforce collectivization,
27–28
Strip farming, 72
Studebaker shoes, 198
Substitution food products, 18
Sugar, availability of, 167–168
Sunflower crops, 37
Surveillance of agricultural
resources. *See* Quarantine
Stations
Svinarki (female swineherds),
137–141, 142*f*
Swine Trust, 137–141, 142*f*

Taezhnik farm, 218, 252*n*4
Taiga landscape, 207, 209
Taylor, Frederick, 7, 17
Taylorism, 17, 120, 121
Tear gas, 87
Territorial Production Complexes
(TPKs), 190, 192, 193, 197,
223
Three Year Plan for Agriculture
(1947), 111, 137–138
Tomatoes, canning of, 178–180
Tractors, 8, 70, 111
Trans-Siberian railway, 190, 196

Trofim Lysenko Station of Animal and Plant Hybridization and Acclimatization. *See* Askaniia Nova breeding station

Trudodni (workday credits), 71–72

Tushonka (canned pork), 151, 152, 156–158, 163–164, 180, 249n4

Twenty-five-thousanders, 63, 73

Ukraine: cattle herds in, 115; collectivization in postwar era, 66; famine in, 3, 70, 74, 75; German invasion and occupation of, 66, 67, 88; postwar reconstruction of, 136; potatoes in, 15, 82, 86, 99, 100; Quarantine Stations in, 84, 91

Ukrainian Grey Cattle, 145, 244n9

Ukrainian White Steppe pig, 133–136, 139, 141, 248n46

United Nations Relief and Rehabilitation Administration, 158

United States: agricultural quarantine facilities in, 84; canned meat shipments to Soviet Union, 156–157; famine, neglect in recognizing, 75–76; farming and livestock operations in, 33, 58–59, 115–116; Foreign Agricultural Service of, 74–75; import ban on goods from Soviet Union, 215; industrialization of animal agriculture in, 112; kolkhozes, perceptions of, 95–96; meat consumption in, 153; perceptions of animal agriculture in Soviet Union, 114–116; propaganda against traitors and agricultural pests, similarities between, 92; revocation of food aid to Soviet Union, 78; wheat sold to Soviet Union, 6, 225–226

Urban bias in distribution networks, 180–181, 186

Vakulin, Gavriil, 96

Vavilov, Nikolai, 123–124

Vegetable-enriched ice cream, 172–173

Vegetables. *See* Fruits and vegetables, canning of

Vernalization, 123

Veterinary field technicians, 29

Violence, as method to enforce collectivization, 27–28

Virgin Lands Campaign, 5, 189, 192–193, 197, 223

Vodka distilleries, 72, 82, 92

Von Thunen rings, 202–205, 206

Vrag naroda (enemy of the people), 91–92

Water contamination, 12

Weed infestations, 72–73, 100

Weiner, Douglas, 132

Western Oblast: map of, 31m; population demographics, 36–37

Wheat: cold hardening of, 123; crop failures, 125; as export commodity, 10; purchase from U.S., 6, 225–226

Wild foods, 175–177

"Will to improve," 34

Windbreaks, 8
Women: on Bratsk dam project,
    198; as milkmaids, 142–143,
    146–147; as pushcart ladies,
    170; as swineherds, 137–141,
    142*f*

Workday credits, 71–72
Workers' cafeterias, 182–183
World War II, impact on Soviet
    Union, 4, 64–65, 67–70

Zemlianki (dugout houses), 80